Remote Sensing and Digital Image Processing with R - Lab Manual

This Lab Manual is a companion to the textbook *Remote Sensing and Digital Image Processing with R*. It covers examples of natural resource data analysis applications including numerous, practical problem-solving exercises, and case studies that use the free and open-source platform R. The intuitive, structural workflow helps students better understand a scientific approach to each case study in the book and learn how to replicate, transplant, and expand the workflow for further exploration with new data, models, and areas of interest.

Features

- Aims to expand theoretical approaches of remote sensing and digital image processing through multidisciplinary applications using R and R packages.
- Engages students in learning theory through hands-on real-life projects.
- All chapters are structured with solved exercises and homework and encourage readers to understand the potential and the limitations of the environments.
- Covers data analysis in the free and open-source R platform, which makes remote sensing accessible to anyone with a computer.
- Explores current trends and developments in remote sensing in homework assignments with data to further explore the use of free multispectral remote sensing data, including very high spatial resolution information.

Undergraduate- and graduate-level students will benefit from the exercises in this Lab Manual, because they are applicable to a variety of subjects including environmental science, agriculture engineering, as well as natural and social sciences. Students will gain a deeper understanding and first-hand experience with remote sensing and digital processing, with a learn-by-doing methodology using applicable examples in natural resources.

Remote Sensing and Digital Image Processing with R - Lab Manual

Marcelo de Carvalho Alves
Luciana Sanches

CRC Press is an imprint of the
Taylor & Francis Group, an **informa** business

Designed cover image: © Shutterstock

First edition published 2023
by CRC Press
6000 Broken Sound Parkway NW, Suite 300, Boca Raton, FL 33487-2742

and by CRC Press
4 Park Square, Milton Park, Abingdon, Oxon, OX14 4RN

CRC Press is an imprint of Taylor & Francis Group, LLC

ISBN: 978-1-032-46124-3 (pbk)
ISBN: 978-1-003-38041-2 (ebk)

DOI: 10.1201/9781003380412

Typeset in Latin Modern font
by KnowledgeWorks Global Ltd.

Publisher's note: This book has been prepared from camera-ready copy provided by the authors.

Contents

About the Authors

Marcelo de Carvalho Alves

Dr. Alves is an associate professor at the Federal University de Lavras, Brazil. His education includes master's, doctoral, and post-doctoral degrees in Agricultural Engineering at Federal University of Lavras, Brazil. He has varied research interests and has published on surveying, remote sensing, geocomputation, and agriculture applications. He has over 20 years of extensive experience in data science, digital image processing, and modeling using multiscale, multidisciplinary, multispectral, and multitemporal concepts applied to different environments. Experimental field sites included a tropical forest, savanna, wetland, and agricultural fields in Brazil. His research has been predominantly funded by CNPq, CAPES, FAPEMIG, and FAPEMAT. Over the years, he has built a large portfolio of research grants, mostly relating to applied and theoretical remote sensing, broadly in the context of vegetation cover, plant diseases, and related impacts of climate change.

Luciana Sanches

Dr. Sanches graduated with a degree in Sanitary Engineering from the Federal University of Mato Grosso, Brazil, a master's degree in Sanitation, Environment, and Water Resources from the Federal University of Minas Gerais, a PhD in Road Engineering, Hydraulic Channels, and Ports from Universidad de Cantabria, Spain, a post-doctorate degree in Environmental Physics, Brazil, and a post-doctorate degree in Environmental Sciences from the University of Reading, United Kingdom. Her education includes postgraduate degrees in Workplace Safety Engineering at Federal University of Mato Grosso, Brazil, and in Project Development and Management for Municipal Water Resources Management at the National Water Agency, Brazil. She is currently an associate professor at the Federal University of Mato Grosso, and worked for more than 20 years in research on atmosphere-biosphere interaction, hydrometeorology in various temporal-spatial scales with interpretation based in environmental modeling and remote sensing. She has been applying remote sensing in teaching and research activities to support the interpretation of environmental dynamics.

Preface

Remote sensing and digital image processing enable us to understand functioning mechanisms and geospatial relationships between ecological variables of agroecosystems and ecosystems, over large areas, repeatedly. This lab manual is a companion to the first edition of the textbook *Remote Sensing and Digital Image Processing with R*. Examples of natural resource data analysis applications are covered including numerous practical, problem-solving exercises, and case studies that use the free and open-source platform R. The book methodology aims to expand theoretical approaches of remote sensing and digital image processing from multidisciplinary applications using R and R packages to engage students in learning theory through hands-on real-life projects. The intuitive, structural workflow helps students better understand a scientific approach to each case study in the book, in order to easily replicate and expand the workflow for further exploration with new data, models, and regions of interest. All chapters are structured with solved exercises and homework and can be used widely in environmental science and agriculture engineering as well as in the physical, natural, and social sciences, at the undergraduate and graduate levels.

The items covered in the chapters are defined in a form of constructive complexity that allows the student to develop a project using remote sensing and geospatial data, and finally, conclude the studies with a report in the form of a scientific article or scientific review of remote sensing applied to agricultural and environmental analysis. The chapters covered are "Principles of R Language in Remote Sensing and Digital Image Processing" (Chapter 1), "Introduction to Remote Sensing and Digital Image Processing with R" (Chapter 2), "Remote Sensing of Electromagnetic Radiation" (Chapter 3), "Remote Sensing Sensors and Satellite Systems" (Chapter 4), "Remote Sensing of Vegetation" (Chapter 5), "Remote Sensing of Water" (Chapter 6), "Remote Sensing of Soils, Rocks, and Geomorphology" (Chapter 7), "Remote Sensing of the Atmosphere" (Chapter 8), "Remote Sensing and Digital Image Processing for Project Design" (Chapter 9), "Visual Interpretation and Enhancement of Remote Sensing Images" (Chapter 10), "Unsupervised Classification of Remote Sensing Images" (Chapter 11), "Supervised Classification of Remote Sensing Images" (Chapter 12), "Uncertainty and Accuracy Analysis in Remote Sensing and Digital Image Processing" (Chapter 13), "Remote Sensing and Digital Image Processing for Article Enhancement" (Chapter 14).

Marcelo de Carvalho Alves and Luciana Sanches, Lavras, June 2023

1

Principles of R Language in Remote Sensing and Digital Image Processing

1.1 Introduction

R is a free software environment for statistical computing and graphics which compiles and runs on a wide variety of UNIX platforms, Windows and MacOS. R is a powerful environment for digital image processing, geocomputation, as well as for obtaining and analyzing remote sensing data (Lovelace et al., 2019). However, R has been used in different areas of knowledge, and applied in teaching different disciplines such as computational statistics (Taylor, 2018), biostatistics (Sarvary, 2014), data mining (Hussain, 2015), data science (Matter, 2021), data analytics (Patil, 2016). R can run on many operating systems, but this book is prepared with R used on a Windows 10 version. Other versions such as Mac and Linux can also be used for R (Douglas et al., 2022; Rodríguez, 2022).

The objective of this chapter is to provide a foundation for readers to quickly move forward with using R in subsequent chapters where there are more advanced applications of processing, analyzing, and geovisualizing geospatial data and remote sensing imagery. To understand and apply digital image processing techniques to remote sensing data, it is essential to know basic objects that store various types of data in R, as well as functions and graphics used to generate scientific knowledge and useful information for thematic mapping.

1.2 The R Language and Environment

The R language was first created and released in a paper published by Ihaka & Gentleman (1996), and is now under active development by a group of statisticians called "**the R core team**[1]". The R language was inspired by the S language developed by John Chambers and others at Bell Labs. However, the modern implementation of R is far more popular than S. R is freely available and distributed under the terms of the Free Software Foundation's GNU General Public License. R software can be obtained from the Comprehensive R Archive Network (CRAN) where R and R packages ready to run on Windows, Mac OS, and Linux are available. The source code is also available for download and can be compiled for other platforms (Rodríguez, 2022).

The popularity of R compared to other computer languages has been prominent among the most used languages in the contemporary world and can be evaluated in relation to its use as a working tool and its use in data science and data analysis software. The creation of new packages has occurred exponentially (Douglas et al., 2022; Hornik, 2012a; Muenchen, 2022).

[1] https://www.r-project.org

The popularity of R can be attributed to factors such as (Douglas et al., 2022):

- R is open source and freely available;
- R is available for Windows, Mac, and Linux operating systems;
- R has an extensive and coherent set of tools for statistical analysis;
- With R it is possible to widely explore graphical aspects with flexibility and the ability to produce publishing-quality science;
- In R there is a growing set of packages freely available to extend R's capabilities;
- R has an extensive online support network and freely available documents.

Furthermore, with the use of R, robust and reproducible research practices can be realized when compared to graphical user interface (GUI) software. Thus, it is possible to write code instead of clicking on links and perform analysis with a permanent and accurate record of all methods used and decisions made during data analysis. The code can be shared along with data so that other scientists, technicians and users can exactly reproduce the analysis. This fulfills one of the principles of open science and reproducible reporting with version control (Douglas et al., 2022).

1.3 Why Learn to Program?

With lower computing costs, lower digital data storage costs, and the spread of the Internet, there has been a marked increase in the availability of digital data describing everyday human activities (Einav & Levin, 2014; Matter, 2021). New business models and economic structures are emerging with use of data as the main commodity, determining technological and economic change related to artificial intelligence. This type of economy relies heavily on the processing, analysis, management of large amounts of digital data, as in the case of remote sensing and digital image processing of the Earth's environment.

The need for proper handling of large amounts of digital data has given rise to the interdisciplinary field of data science, as well as to a growing demand for data scientists. This type of science approach is related to the scientific process of remote sensing, where responses to challenges are needed for monitoring and problem solving using a combination of skills and knowledge from different areas of knowledge, including geomatics, geocomputation, geostatistics and statistics.

Remote sensing and digital image processing with R can be used in conjunction with recent developments in data science and career opportunities related to machine learning and programming, bringing benefits and scientific and technical quality to undergraduate, graduate, and service work in areas related to geographic problem solving, agricultural and environmental management, and forensics.

R is a high-level computer language, relatively easy to learn for people with no previous programming experience. The syntax is quite intuitive, with few complex error messages.

Thus, learning by doing a routine previously prepared by the teacher can facilitate learning. Also, with the recent sharp increase in popularity of R, there are many freely accessible online resources that help beginners learn and use the language (Matter, 2021).

1.4 History of R

R is an open source implementation of the S programming language that allows you to define objects in predetermined blocks instead of the entire code. S was introduced by Becker & Chambers (1984). Since then, new versions of the S language have been created. In 1994, Ross Ihaka & Gentleman (1996) wrote a version of S at the University of Auckland and named it R (Patil, 2016). In 1995, R software was made free and open source under the GNU General Public License. The first official version was released in June 1995. In 1997 R became an official part of the GNU collaborative free software project, with code hosted and maintained by the Concurrent Versions System (CVS). The Comprehensive R Archive Network (CRAN) was officially founded in 1997 (Hornik, 2012b). The R Core Team was formed in 1997 to further develop the R language. In 1999 the first versions of functions for downloading, installing and updating packages appeared. The first official stable version of R for production and use was released on February 29, 2000 (Dalgaard, 2008; Ihaka, 1998). In April 2003, the R Foundation was designated as a non-profit organization to provide further support for the R project (Fox, 2009). Since January 2022, it consists of the R Core team, including collaborative results from scientists around the world (R Core Team, 2020).

In CRAN are stored R executable files, source code, documentation, and user-contributed packages. CRAN originally had 3 mirrors and 12 packages. Since January 2022, there are 101 mirrors and 18,728 contributed and hosted packages of programs (binaries) for major Linux, macOS, and Windows distributions (R Core Team, 2020; The R Foundation, 2022).

Currently there is an extensive and rapidly growing R literature. In the area of remote sensing and digital image processing of remote sensing data, much of the existing literature and examples is the work of package developers and R users around the world, with emphasis on development-related scientists, operation and updating of packages such as `raster` (Hijmans et al., 2020), `terra` (Hijmans et al., 2022), `luna` (Ghosh et al., 2021), `stars` (Pebesma et al., 2022), `sf` (Pebesma et al., 2021), `RStoolbox` (Leutner et al., 2019), `rgee` (Aybar et al., 2022), `ggplot2` (Wickham et al., 2022), `tmap` (Tennekes et al., 2020), `mapview` (Appelhans et al., 2020), and others.

The open source R programming language and statistical computing environment has in the last decade become a key tool for data science in industry and academia. Its initial conception was for statistical analysis, but nowadays the potential for remote sensing data analysis has been expanded with the update and emergence of new R packages to handle large volumes of data with efficient processing algorithms. Many features of the language have made R useful for working with data, and it has been extended with the growth of data economics and data science, with no limit on its use in various domains, going these days far beyond traditional academic research applications (Matter, 2021).

Official R manuals are distributed as PDF files or online help service that can assist on introductory and advanced aspects from installation to data manipulation and dissemination of results.

1.5 R's Growth

Advances in technology have paved the way for increasingly powerful, sophisticated, and comprehensive data analysis tools. Programming languages allow analysts to define or customize their own functions or apply functions developed by experts around the world. By working directly with programming languages, analysts can design and implement operations, models, and other tools that meet specific needs.

Programming languages can be ranked in terms of demand and popularity of programmers. Before choosing a programming language, one should consider factors such as popularity, demand, career opportunities, and applications. In 2015, R was ranked sixth out of 10 languages cited. Furthermore, as the amount of data-intensive work increases, the demand for tools such as R for data mining, processing, and visualization also tends to increase (Patil, 2016). Currently, R has been cited among 10 programming languages as a recommendation for 2022 (Abbott, 2021).

1.5.1 R in business

R originated as an open-source version of the S programming language in the 1990s. It has since gained the support of several companies, most notably RStudio and Revolution Analytics, which are used to create packages and services related to the language. R is backed by large companies that power some of the largest relational databases in the world such as Oracle (Patil, 2016).

Many large technology companies like Uber, Google and Facebook use R programming language for business management. With the growing demand for machine learning and data science, the R programming language is worth learning (Abbott, 2021). Ford, Twitter, US National Weather Service, the Rockefeller Institute of Government, and the Human Rights Data Analysis Group are cited as users of the R language (Patil, 2016).

1.5.2 R in higher education

R originated in academia, by Ross Ihaka and Robert Gentleman at the University of Auckland, New Zealand. From there, R has also been widely adopted in graduate programs that include intensive study of statistics, such as Coursera Data Science Program (Patil, 2016).

Some reasons for which educational institutions are using R include:

- R is free, saving institutions from expensive software licenses and students from unnecessary costs;
- R gives students career-ready data science skills;
- R makes reproducible research the norm;
- R makes it possible to visualize results;
- R has broad compatibility and can be used for various types of data, including large amounts of data.

1.5.3 R community

Considering the vast number of contributors, R serves as a valuable tool for users in different areas related to data analysis and computation. R users have direct access to thousands of freely available packages covering various aspects of data analysis, statistics, data elaboration and data import. Many users apply R as a daily work tool without necessarily writing programs, but using packages and routines written in R to solve simple and complex problems.

Almost all functions that a modern commercial computing environment with a focus on statistics offers can be found within the R environment. Furthermore, there are R packages that cover all areas of modern data analysis, including natural language processing, machine learning, and big data analysis. Thus, there is really no need to write a program for many tasks performed with R, so an analysis routine can be created from existing packages and with very high quality (Matter, 2021).

There is an extensive community around R, and abundant courses, tutorials, books, blogs, websites, freely available online for consultation. A small list of links are made available for beginners to consult as introductory support for using R (Marques, 2019):

- R webpage[2] <- The main R webpage, with links to R downloads, manuals, tutorials, information search system about the R project;
- R video tutorials[3] <- A video introduction for beginners in R;
- Online course[4] <- A short course in R;
- Short reference card[5] <- R reference card with handy list of useful R functions;
- Long reference card[6] <- Reference card with frequently used R functions;
- RStudio cheatsheets[7] <- Cheatsheets that make it easy to use some favorite packages in R.

1.5.4 R is powerful

R is powerful, with the ability to generate maps, charts, and data analysis results in very few lines of code and with optimized processing. Tasks that require multiple lines of code can be accomplished in R with few lines and processing large amounts of data (Patil, 2016).

1.6 Benefits of Using R

Due to its wide range of contributors, R serves as a valuable tool for users in various fields related to data analysis and computation associated with the package ecosystem, extensibility, open source, graphical and mapping capabilities and strong connection with academia (Patil, 2016).

1.6.1 Package ecosystem

One of the strongest qualities of R is the abundance of the package ecosystem. There is a lot of functionality that is built in for statisticians, engineers, and scientists in general (Patil, 2016).

1.6.2 R is extensible

R provides rich functionality for developers to build their own tools and methods for data analysis. A large number of people have emerged from other fields such as biological sciences and humanities. People can extend it without asking permission (Patil, 2016).

[2] http://r-project.org/
[3] http://blog.revolutionanalytics.com/2013/08/google-video-r-tutorials.html
[4] http://faculty.washington.edu/tlumley/Rcourse/
[5] http://cran.r-project.org/doc/contrib/refcard.pdf
[6] https://cran.r-project.org/doc/contrib/Short-refcard.pdf
[7] https://www.rstudio.com/resources/cheatsheets/

1.6.3 Free software

At the time R first appeared, its biggest advantage was that it was free software. Everything there was free and its source code was available (Patil, 2016).

1.6.4 R's graphics and charting capabilities

Packages like `dplyr` (Wickham et al., 2021), `ggplot2` (Wickham et al., 2022) literally improved the quality of life for data manipulation and plotting, respectively (Patil, 2016). Currently, this list is large and different package options will be presented throughout the book.

1.6.5 R's strong ties to academia

Any new line of research probably has an associated R package. So R continues to be progressive. The `caret` package (Kuhn et al., 2020) offers a very clever form of machine learning in R with many popular algorithms already implemented (Patil, 2016).

1.7 R's Challenges

Despite all the benefits, R can have limitations in memory management, speed, and efficiency as probably the biggest challenges faced. In addition, there can be problems when working with very large datasets. However, some problems can be solved by using third-party software and other languages associated with R. Although an analysis can be done in R, the delivery of results can be done in a different language such as JavaScript (Patil, 2016).

1.8 Big Data Strategies in R

The R language has become popular for being interactive and elevating research with clarity and display. It enables scientists to quickly and repeatably process big data to create a powerful and reliable statistical model, transform data, evaluate multiple model options, and visualize results (Patil, 2016).

1.9 Installation and Interface

A basic presentation on how to obtain and install R and RStudio on your computer is presented with brief guidance on using R in the command console, installing and working with R packages to extend R's capabilities, habits for working on projects, documenting workflow, and writing readable R code (Douglas et al., 2022).

1.9.1 Installing R and RStudio

The first step is to install R. R is freely available for Windows, Mac, and Linux operating systems from the Comprehensive R Archive Network (CRAN) **website**[8]. For Windows and Mac users, we suggest downloading and installing the precompiled binary versions.

While it is possible to use R from its basic installation, we will use a popular Integrated Development Environment (IDE) called "RStudio".

RStudio is an add-on to R with a more user-friendly interface, incorporating the R Console, a script editor, and other useful features (such as R markdown and GitHub integration). More information about RStudio can be found **here**[9].

RStudio is freely available for Windows, Mac and Linux operating systems and can be downloaded from the **RStudio website**[10]. It is recommended that you choose the RStudio Desktop version. You must install R before installing RStudio.

See this **video**[11] for step-by-step instructions on how to download and install R and RStudio.

A basic differentiator of RStudio over R, in the console, is that the commands are autocompleteable (Sanatan, 2017).

Some aspects of RStudio are configured in the `Tools -> Global options` menu, including the version of R that will be used as a reference by RStudio.

1.9.2 Installing and loading packages

Basic statistical functions are already programmed to work from the basic installation of R. However, more advanced functions developed by the scientific community in general require the use of packages applied to specific situations. Once R is installed, a package needs to be installed only once. However, a package needs to be loaded every time R is used in an application that requires the specific use of that package. The operation of a package may also depend on other packages installed in R for full operation or operation of all functions available in a package. Packages can be installed via the R or RStudio menu, as well as by the `install.packages` function. Packages are enabled with the `library` or `require` functions.

For performing some homework tasks, it is necessary to use cloud computing via `Earth Engine`. In this case, the `rgee` package and dependencies in R, Python and Google account are required for full operation of these activities. Information for installing `rgee` can be obtained from the **vignettes**[12] provided on CRAN where the package installer is obtained. Videos on **YouTube**[13] can also be accessed as support for installing `rgee` and dependent packages.

[8] https://cloud.r-project.org/
[9] https://rstudio.com/
[10] https://rstudio.com/products/rstudio/download
[11] https://bit.ly/3cuRqSZ
[12] https://cran.r-project.org/web/packages/rgee/index.html
[13] https://www.youtube.com/watch?v=_fDhRL_LBdQ

In the case of the `SegOptim` package, it is also necessary to install **third-party software**[14] depending on the object-based segmentation used.

1.9.3 R and RStudio interface

R is an easy-to-use programming language initially developed for statistical computing. Once R is installed on our computer, we can run it directly from the command line by typing R and pressing enter, with the Integrated Development Environment (IDE) with the basic R installation, or with the more elaborate and user-friendly IDE called "RStudio local" or "in the cloud". RStudio is a very useful tool for simple data analysis with R, for writing R scripts (small R programs), or even for developing R packages (software written in R), as well as for building interactive documents, presentations, etc. In addition, it offers many options to change its own appearance (Pane Layout, Code Highlighting, etc.)

In the following, we address each of the main panels that will be relevant in practice activities (Figure 1.1).

RStudio's interface is split so that on the left side is the console, where commands can be typed and where the results of executed commands are displayed. On the upper right side there are two tabs, `Environment`, where the created objects and imported databases are stored, and `History`, where the history of executed commands is stored. The `Environment` tab shows the objects that have been created. On the lower right side there are 5 tabs. The `Files` tab shows the files in the working directory. In the `Plots` tab are the generated plots, in the `Packages` tab are listed the installed packages, and the `View` tab for viewing web content (Sanatan, 2017).

1.9.3.1 R console

When working in an interactive session, you simply type R commands directly into the R console. Usually, the output of executing a command in this way is also printed directly on the console. So we type a command on one line, click enter, and the output is displayed on the next line.

For example, we can tell R to print the phrase "remote sensing and digital image processing with R" on the console by typing to follow the command on the console and pressing enter using the "print" function.

```
print("remote sensing and digital image processing with R")
#[1] "remote sensing and digital image processing with R"
```

1.9.3.2 Script editor

Apart from very short interactive sessions, it usually makes sense to write R-code not directly on the command line, but in an R-script in the script panel. This way you can easily run several lines at the same time, comment out the code with explanations, register it on your hard disk, and continue developing the code later.

[14] https://segoptim.bitbucket.io/docs/installation.html#installing-third-party-software-for-image-segmentation

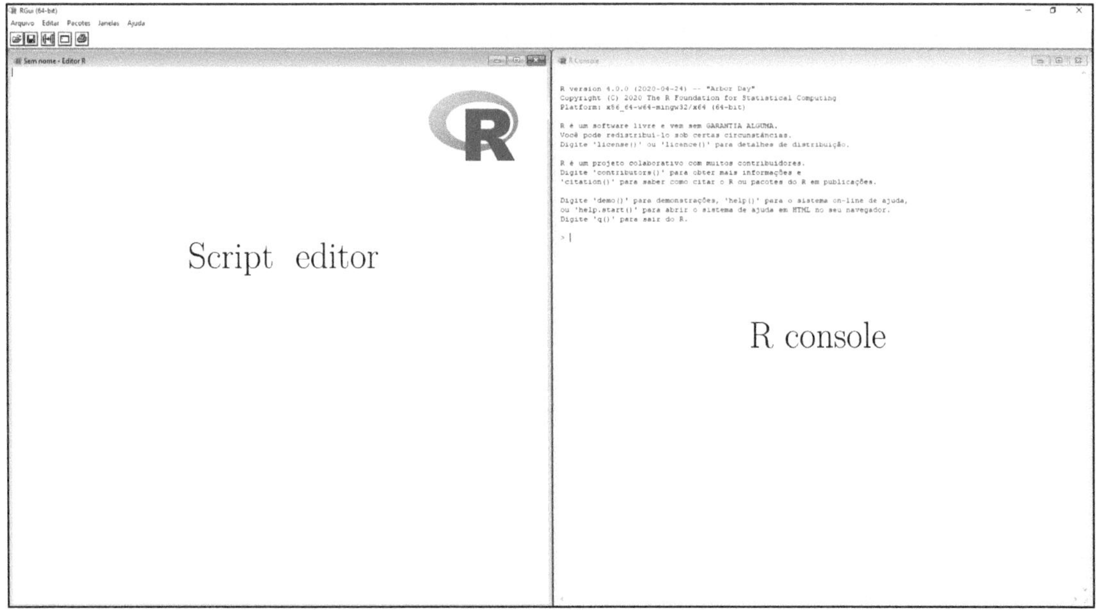

FIGURE 1.1 R 4.0 and RStudio 1.3.959 interface with windows.

1.9.3.3 Environment explorer

The environment pane shows which variables, objects, and data are loaded in a current R session. In addition there are functions for opening documents and importing data.

1.9.3.4 File explorer, help and viewer

With the file explorer window it is possible to browse through the structure of folders and files on our computer's hard disk, modify files, and set the working directory of our current R session. In addition, there is a window to display graphs generated in R and a window with help pages and R documentation.

The `plot` function is used as an example of creating a graph in RStudio (Figure 1.2).

```
plot(1:100)
```

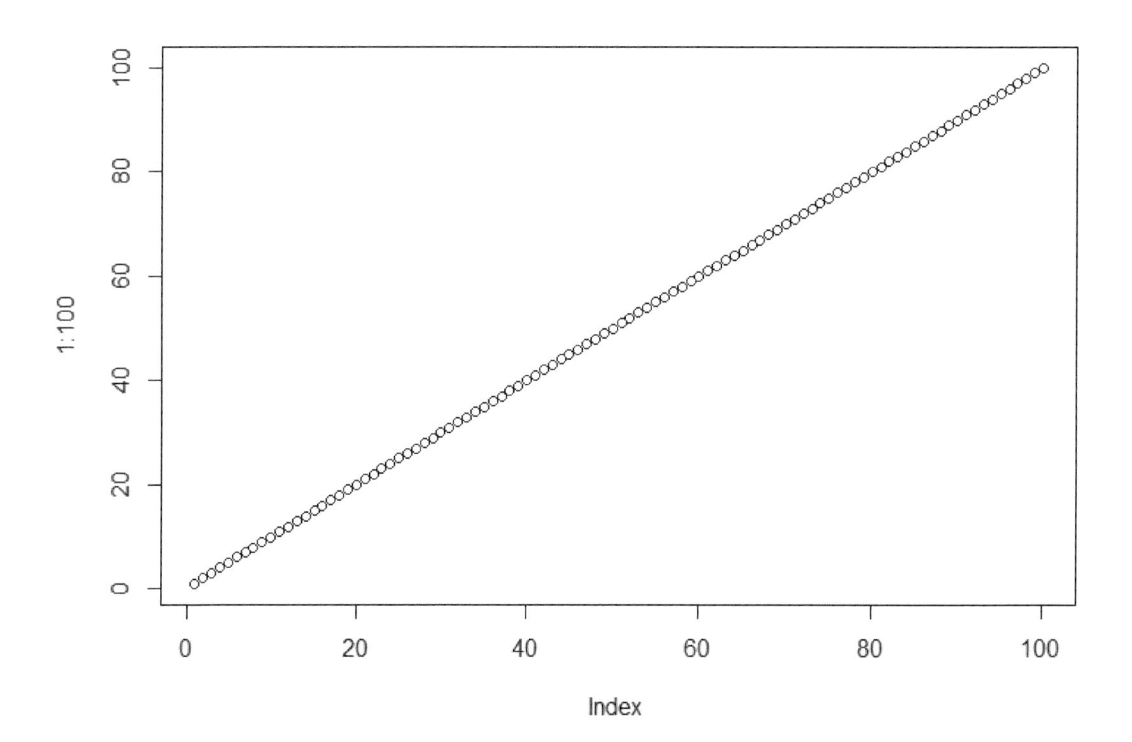

FIGURE 1.2 Creating graphics in RStudio.

1.9.4 Dynamic reports and reproducible research

One of the most amazing features of the integration between R and RStudio is the possibility to work with dynamic reports in a simple way. Thus, manuscripts, books, and presentations can be prepared via R and RStudio. For this, you need to know the basics of **R Markdown**[15]. More details about **R Markdown**[16] on the Internet. A complete book on **R Markdown**[17] can be read on this same topic from a GitBook available on the Internet (Marques, 2019; Xie et al., 2018).

1.10 R Basics

Open RStudio. The default is that an empty workspace appears. If you have an existing workspace, you can open it by selecting `File -> Open File`. We recommend that you start by creating a script file (`Ctrl+Shift+N`, shortcut RStudio) and use it to save and comment out all your code that will be run during the tutorial to record everything that has been done.

The console in RStudio (or R) is where you type R expressions and see the results of the text. For example, when you exit RStudio (and R) just type `q()`.

```
q()
```

Notice the parentheses after `q`. This is because in R you do not type commands, but rather functions are called to achieve results, even to exit the software. To call a function, type the function name followed by the arguments in parentheses. If the function does not take arguments, just type the name followed by the left and right parentheses. After you type `q()`, you are asked if you want to save the current working environment. This prompt can be ignored by typing `q("no")`. It is also important to know the `help()` function, which uses the bottom right window (opens a separate help window in R). The function can be called with arguments to get help on specific features for using and understanding functions in R. A shortcut to help on a topic is a question mark followed by the topic, as in `?plot`.

```
help(plot) # Help
?plot # Help
```

The up and down arrow keys can be used to scroll through recent lines of code input. So if you make a mistake, all you need to do is hit the up key on the console to recall the last expression and edit it correctly. You can also write a script in the top window of the RStudio console and then click `Run` in the top right corner of the screen to run the script.

1.10.1 Using R as a calculator

R can be used as a calculator by typing the mathematical expression of interest or an existing function into the console to obtain the result of a mathematical function. The default arithmetic operators are `+`,`-`,`*`, and `/` for adding, subtracting, multiplying, and dividing, respectively. The `^` operator is used for exponentiation. Mathematical functions such as `sqrt()`, `exp()`, and `log()` can be used. R follows the usual mathematical convention on the order of operations so that the bracket operator can change the result of calculations.

[15] https://rmarkdown.rstudio.com/authoring_basics.html
[16] https://rmarkdown.rstudio.com/
[17] https://bookdown.org/yihui/rmarkdown/

```
1+1 # arithmetic expression
#[1] 2
2+3*4 # arithmetic expression
#[1] 14
(2+3)*4 # arithmetic expression
#[1] 20
3^4 # 3 to the power of 4
#[1] 81
log(10) # logarithm to base e
#[1] 2.302585
log10(10) # logarithm to base 10
#[1] 1
exp(10) # natural antilog
#[1] 22026.47
sqrt(10) # square root
#[1] 3.162278
pi # pi
#[1] 3.141593
```

Functions and operators are grouped into three categories and grouping methods can be written for each of these categories. There is no mechanism yet for adding groups. You can write specific methods for any function within a group. Functions for different groups are listed below (Table 1.1).

TABLE 1.1 Functions used for different groups.

Group	Function
Math	abs, acos, acosh, asin, asinh, atan, atanh, ceiling, cos, cosh, cospi, cumsum, exp, floor, gamma, lgamma, log, log10, round, signif, sin, sinh, sinpi, tan, tanh, tanpi, trunc
Summary	all, any, max, min, prod, range, sum
Ops	+, -, *, /, ^, < , >, <=, >=, !=, ==, %%, %/%, &, |, !

1.10.2 Comparing values and getting logical answers

R also understands the relational operators `<=`, `<`, `==`, `>=`, `>=` and `!=` for less than or equal, less than, equal, greater than, greater than or equal, and not equal, respectively. You can use them to create logical expressions that accept TRUE (`T`) and FALSE (`F`) values as output. Logical expressions can be combined with logical operators.

```
3*3 == 9 # Logical operation
#[1] TRUE
4*4 >= 15 # Logical operation
#[1] TRUE
a1 = c(TRUE, TRUE, FALSE, FALSE) # Creating a1
a2 = c(TRUE, TRUE, TRUE, FALSE) # Creating a2
a1 & a2 # a1 e a2
#[1]  TRUE  TRUE FALSE FALSE
```

Note that when you execute the code as above, the result is displayed in the console. As an option, you can store the results in an object; this can be useful in other situations.

1.10.3 Creating variables by assigning values

In general, objects can have classes, which allow functions to interact with them. Objects can be of various classes. We have already used the numeric class, which is used for general numbers, but there are also additional classes that are very commonly used:

- `integer` <- Integer numbers;
- `character` <- Character strings;
- `factor` <- Levels of a categorical variable;
- `logical` <- TRUE and FALSE values.

1.10.4 Evaluating and assigning data structures

It is possible to check the type or class of any object using the `class` function.

```
numb <- 10
class(numb)
#[1] "numeric"
chara <- "class"
class(chara)
#[1] "character"
logic <- TRUE
class(logic)
#[1] "logical"
```

Alternatively, you can ask whether an object is of a specific class, using a logical test associated with the `is.` code.

```
is.numeric(numb)
#[1] TRUE
is.character(numb)
#[1] FALSE
is.character(chara)
#[1] TRUE
is.logical(logic)
#[1] TRUE
```

Sometimes it can be useful to change the class of a variable using the `as.` code associated with the class name of interest. In this case, the previously created numeric object is converted into the class `character`.

```
class(numb)
#[1] "numeric"
num_chara <- as.character(numb)
num_chara
```

```
#[1] "10"
class(num_chara)
#[1] "character"
```

1.10.5 Creating and working with vectors

R is also designed to work with vectors. The `c()` function, which is short for catenate, or concatenate if you prefer, can be used to create vectors from scalars or other vectors.

```
# Create x vector
x <- c(10, 20, 30, 40, 50, 60, 70, 80, 90, 100)
# Calculation with vectors
x+5
#[1]  15  25  35  45  55  65  75  85  95 105
# Create y vector
y <- c(1, 2, 3, 4, 5, 6, 7, 8, 9, 10)
# Calculation between vectors
x+y
#[1]  11  22  33  44  55  66  77  88  99 110
# New vectors using past vectors
z <- c(x,y)
z
#[1] 10 20 30 40 50  60  70  80  90 100   1   2   3   4   5   6   7   8   9  10
# Third value of the x vector
x[3]
#[1] 30
# First value of the y vector
y[1]
#[1] 1
# Add first value of vector x and fifth value of vector y
x[3] + y[1]
#[1] 31
# Take values first to fifth from x vector
x[1:5]
#[1] 10 20 30 40 50
```

1.10.6 Matrices

With R you can also use matrices and arrays of higher dimension. A 4 by 4 matrix is created below with numbers from 1 to 16.

```
m <- matrix(1:16, 4, 4)
m
#      [,1] [,2] [,3] [,4]
#[1,]     1    5    9   13
#[2,]     2    6   10   14
```

```
#[3,]    3    7   11   15
#[4,]    4    8   12   16
```

1.10.7 Using statistics

Statistical functions can be applied to previously created vectors in order to perform a variety of operations.

```
# Sum all values in x vector
sum(x)
#[1] 550
# Take the mean and median for all x values
mean(x)
#[1] 55
median(x)
#[1] 55
# Obtain summary statistics for z vector
summary(z)
#Min. 1st Qu.  Median    Mean 3rd Qu.    Max.
#1.00    5.75   10.00   30.25   52.50  100.00
# Repeat number five, ten times
rep(5, times = 10)
#[1] 5 5 5 5 5 5 5 5 5 5
```

1.10.8 Creating data frames and tibbles

1.10.8.1 Data frame

A data frame is created from the `yield`, `area`, `crop` objects using the `data.frame` function. Using the `str` function gives a compact summary of the structure of the data frame object created.

```
yield <- c(180, 155, 160, 167, 181)
area <- c(65, 50, 52, 58, 70)
crop <- c("maize_a", "maize_b", "maize_c", "maize_d", "maize_e")
dataf <- data.frame(prod = yield, area_ha = area, cropland = crop,
stringsAsFactors = TRUE)
str(dataf)
#'data.frame':  5 obs. of  3 variables:
# $ prod     : num  180 155 160 167 181
# $ area_ha : num  65 50 52 58 70
# $ cropland: Factor w/ 5 levels "maize_a","maize_b",..: 1 2 3 4 5
```

1.10.8.2 Tibble

One of the goals of `tibble` is to be a more organized frame structure than those implemented in older R classes, such as `data.frame`. For example, in `tibble`, no names are assigned to the lines. A `tibble` fits the size of your command prompt and prints only the first few rows of the data set. By looking at a tibble you can get a good understanding of what the data structure is, with the data type and variables. The `tibble` package is enabled and then the `as_tibble` function is used to convert the data frame created earlier into a `tibble`.

```
library(tibble) # Enable the package
as_tibble(dataf) # Convert to tibble
# A tibble: 5 x 3
#    prod area_ha cropland
#   <dbl>   <dbl> <fct>
#1    180      65 maize_a
#2    155      50 maize_b
#3    160      52 maize_c
#4    167      58 maize_d
#5    181      70 maize_e
```

1.10.9 Checking and configuring a data directory

The data directory can be configured in some situations where it is necessary to work with files located within a specific folder on the computer. The `setwd` function is used to set up the working directory in a custom location, while the `getwd` function makes it possible to check the directory being used.

```
getwd() # Check working directory
#[1] "C:/bookdown/SRcomR_Lab_Manual"
setwd("C:/bookdown/SRcomR_Lab_Manual") # Set working directory
```

1.10.10 Simple graphics

R has extensive and powerful graphing capabilities. In this example, the `seq` function is used to create equally spaced points between -3 and 3 at intervals of 0.1. The `dnorm` function is used to calculate the standard normal density evaluated at these points. A plot of the normal curve is made with the `plot` function and argument `type = "l"` to plot the curve with lines and not points (Figure 1.3). The `title` function can be used to specify a title on the graph.

```
z = seq(-3, 3, 0.1)
d = dnorm(z)
plot(z, d, type = "l")
```

Another example graph can be used to demonstrate R's use of color. The `pie` function is used to create a graph with 64 slices. The slices are all the same width, the filling is done with different colors obtained using the `rainbow` function (Figure 1.4).

```
par(mfrow = c(1,1), mar = c(1, 1, 0, 0), mgp = c(1.5, 0.6, 0))
pie(rep(1, 64), col = rainbow(64))
```

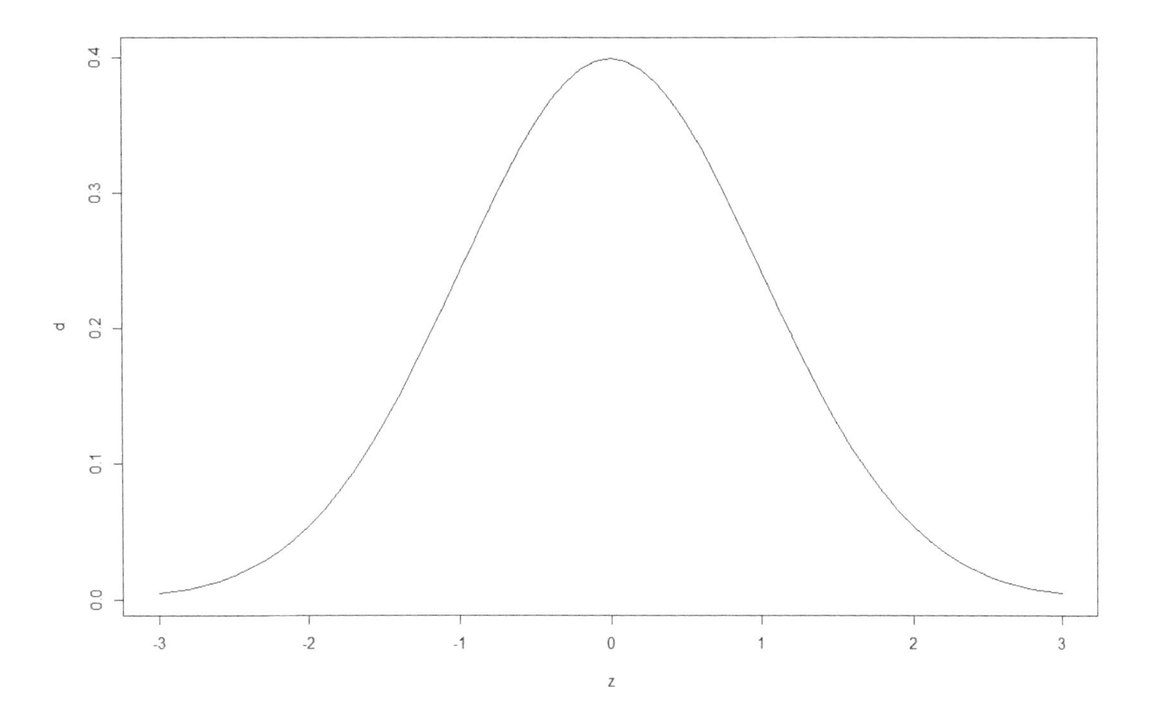

FIGURE 1.3 Standard normal density curve.

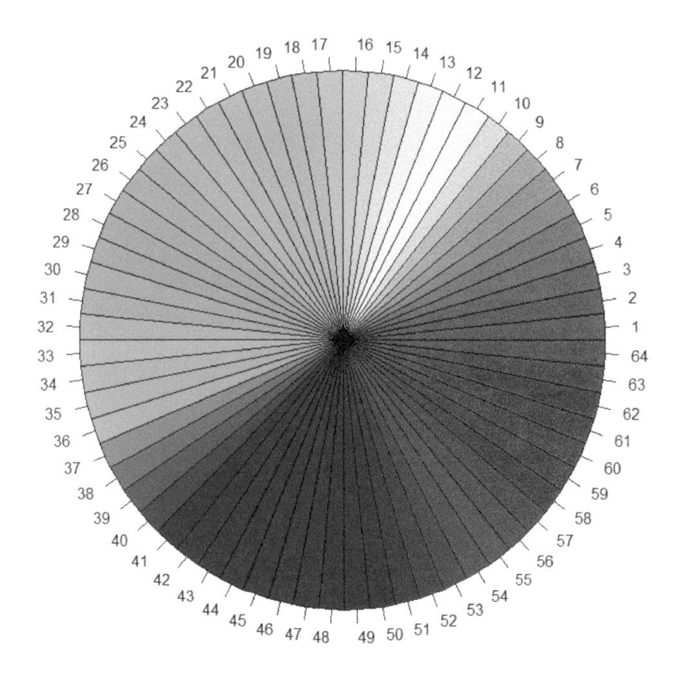

FIGURE 1.4 Slice chart used to display rainbow colors in R.

1.10.11 Citing R

The `citation()` function is used to get the citation and publication information for `R` in `BibTex` format.

```
citation()
#To cite R in publications use:
#R Core Team (2020). R: A language and environment for statistical computing.
#R Foundation for Statistical Computing,
#  Vienna, Austria. URL https://www.R-project.org/.
#A BibTeX entry for LaTeX users is
#  @Manual{,
#    title = {R: A Language and Environment for Statistical Computing},
#    author = {{R Core Team}},
#    organization = {R Foundation for Statistical Computing},
#    address = {Vienna, Austria},
#    year = {2020},
#    url = {https://www.R-project.org/},
#  }
#We have invested a lot of time and effort in creating R,
#please cite it when using it for data analysis. See also
#'citation("pkgname")' for citing R packages.
```

The `package` parameter can be used to get the bibliographic reference of a given package.

```
citation(package = "luna")
#To cite package 'terra' in publications use:
# Robert J. Hijmans (2022). terra: Spatial Data Analysis. R package version 1.5-21.
#  https://CRAN.R-project.org/package=terra
#A BibTeX entry for LaTeX users is
#  @Manual{,
#    title = {terra: Spatial Data Analysis},
#    author = {Robert J. Hijmans},
#    year = {2022},
#    note = {R package version 1.5-21},
#    url = {https://CRAN.R-project.org/package=terra},
#  }
```

1.11 Solved Exercises

1.11.1 Create a working directory with RStudio.

A:

1. Open RStudio and get familiar with the file browser pane on the lower right. Navigate to a folder on your hard disk in which you want to work and store all the code you write.

2. Use the `New Folder` button to create a new folder and name this new folder.

3. You should see the new folder listed in the file browser. Click on it to navigate to its contents. Now, click on the `More` button and select `Set as Working Directory` in the drop-down menu.

1.11.2 Create a script with R.

A:

1. In the RStudio menu bar, select `File/New File/R Script` to create a new file, `shown/opened` in the Script pane.

1.11.3 Examine the different components of the help file.

A: Different methods can be used to get help.

```
help(mean)
?mean
help("mean")
```

1.11.4 Create numeric and character variables and assign it a value.

A: The value `29` is created as a numeric object and the term `remote sensing` as the first character.

```
first_num <- 29
first_char <- "remote sensing"
```

1.11.5 Remove one object and list all objects created in your environment.

A: The `rm` function is used to remove the `first_num` object `ls` is used to list the objects created in the environment.

```
rm(first_num)
ls()
#[1] "first_char"
```

2

Introduction to Remote Sensing and Digital Image Processing with R

2.1 Homework Solution

2.1.1 Subject

Definition of geographic region of interest for use in remote sensing.

2.1.2 Abstract

The use of vector features in remote sensing and image processing can be of primary importance in defining a specific region to be studied. Other conventional functions of using vector geometries of point, line, polygon and corresponding multiples of these geometries can be used as support for obtaining, extracting, editing, viewing and manipulating images in geographic information systems. Different ways of creating vector data are presented, ranging from simpler methods of vector editing on images in Google Earth, to hybrid vector data processing of geometry associated with databases in different classes. These classes have interoperability with each other and can be stored on the computer as well as in the cloud, using R and Earth Engine connected in a multi-purpose processing environment for processing and generating geographic information products. With the type of data processing demonstrated, it is possible to define not only political geometries of borders established on the territory, but also points located in remote locations on the Earth's surface, such as inside the Amazon Rainforest.

Keywords: Earth engine, geographic location, vector data, mapping, `rgee`.

2.1.3 Introduction

Vector features with geometries of points, lines, and polygons can be used to support the identification, extraction of information, and mapping of remotely sensed targets. Vector features can be created with vector geometric information and more complex structures can be obtained by adding a reference coordinate system (CRS) and geographic database with attributes of the object. Different vector data formats can be created and transformed into different classes of files used in remote sensing image processing.

2.1.4 Objective

The objective of the first homework is to create a polygon that defines an area of interest that can be used in remote sensing applications. The polygon can be created on Google Earth by making a drawing on a high spatial resolution remote sensing image. Thus, it is possible to observe details of the area where the polygon is created. After creating the polygon in Google Earth, configuring a closed geometric figure, it can be saved, imported into R, and mapped to make a first map. In this case, a geographic region different from the one presented in the teacher examples should be chosen in order to evaluate learning with another similar application.

2.1.5 Define geographic region from a polygon drawn in Google Earth

A polygon created in Google Earth with the `KML` extension can be imported into R. This vector data is relative to an object or remote sensing process defined over some place on Earth by creating a polygon drawn on top of the target of interest, as in a satellite image color composite.

2.1.6 Mapping `KML` polygons in R using the `sf` and `mapview` packages

The R software is used as an example; however, other software can be used to perform this task. The `sf` (Pebesma et al., 2021), `dply` (Wickham et al., 2021), and `mapview` (Appelhans et al., 2020) packages are needed for this work, and they are loaded with the `library` function.

```
library(sf)
library(dplyr)
library(mapview)
```

2.1.7 Importing polygons of interest in analysis

Polygons of interest are created for analysis at specific locations in the Planet mosaic at the location of interest. A rectangular polygon around the Bom Jesus farm and an indigenous land are used as a reference for locating the areas. The `st_read` function is used to import the polygons in `KML` format into R.

```
ti <- st_read("C:/sr/S2/Trilha0020.kml") # Indigenous land
#Reading layer Layer #0' from data source C:\sr\S2\Trilha0020.kml' using driver KML'
#Simple feature collection with 1 feature and 2 fields
#Geometry type: POLYGON
#Dimension:     XYZ
#Bounding box:  xmin: -47.01409 ymin: -2.883107 xmax: -46.41725 ymax: -1.760037
#z_range:       zmin: 0 zmax: 0
#Geodetic CRS:  WGS 84
fbj <- st_read("C:/sr/S2/bjkml.kml") # Bom Jesus farm
#Reading layer bjkml' from data source C:\sr\S2\bjkml.kml' using driver KML'
#Simple feature collection with 1 feature and 2 fields
#Geometry type: POLYGON
#Dimension:     XYZ
#Bounding box:  xmin: -46.82562 ymin: -1.838801 xmax: -46.70741 ymax: -1.662143
```

```
#z_range:       zmin: 0 zmax: 0
#Geodetic CRS:  WGS 84
```

2.1.8 Remove the Z dimension

The Z dimension observed in the file head is not needed for the mapping. The `st_zm` function is used with the `drop=TRUE` argument in order to remove the Z dimension from the file, wich is necessary for further mapping.

```
tizm<-st_zm(ti, drop = TRUE)
head(tizm)
#Simple feature collection with 1 feature and 2 fields
#Geometry type: POLYGON
#Dimension:     XY
#Bounding box:  xmin: -47.01409 ymin: -2.883107 xmax: -46.41725 ymax: -1.760037
#Geodetic CRS:  WGS 84
#         Name Description                            geometry
#1 Trilha 0020             POLYGON ((-46.87984 -1.8406...
bjzm<-st_zm(fbj, drop = TRUE)
head(bjzm)
#Simple feature collection with 1 feature and 2 fields
#Geometry type: POLYGON
#Dimension:     XY
#Bounding box:  xmin: -46.82562 ymin: -1.838801 xmax: -46.70741 ymax: -1.662143
#Geodetic CRS:  WGS 84
#  Name Description                            geometry
#1                  POLYGON ((-46.75901 -1.6621...
```

2.1.9 Mapping the polygons

The `mapview` function is used to map the Alto Rio Guamá indigenous land and the Bom Jesus farm, state of Pará, Brazil. A template from the `OpenStreetMap` database is used as an associated geovisualization option for the mapping performed (Figure 2.1).

```
mapview(list(tizm,bjzm), col.regions=list("red", "blue"), col=list("black","black"))
```

2.1.10 Converting the polygons to a single file

The polygons are converted into a single file containing the simple polygon features with attributes. The `bind_rows` function of the `dplyr` package is used to list all polygons in a single `sf` file.

```
single_sf <- dplyr::bind_rows(list(tizm,bjzm))
```

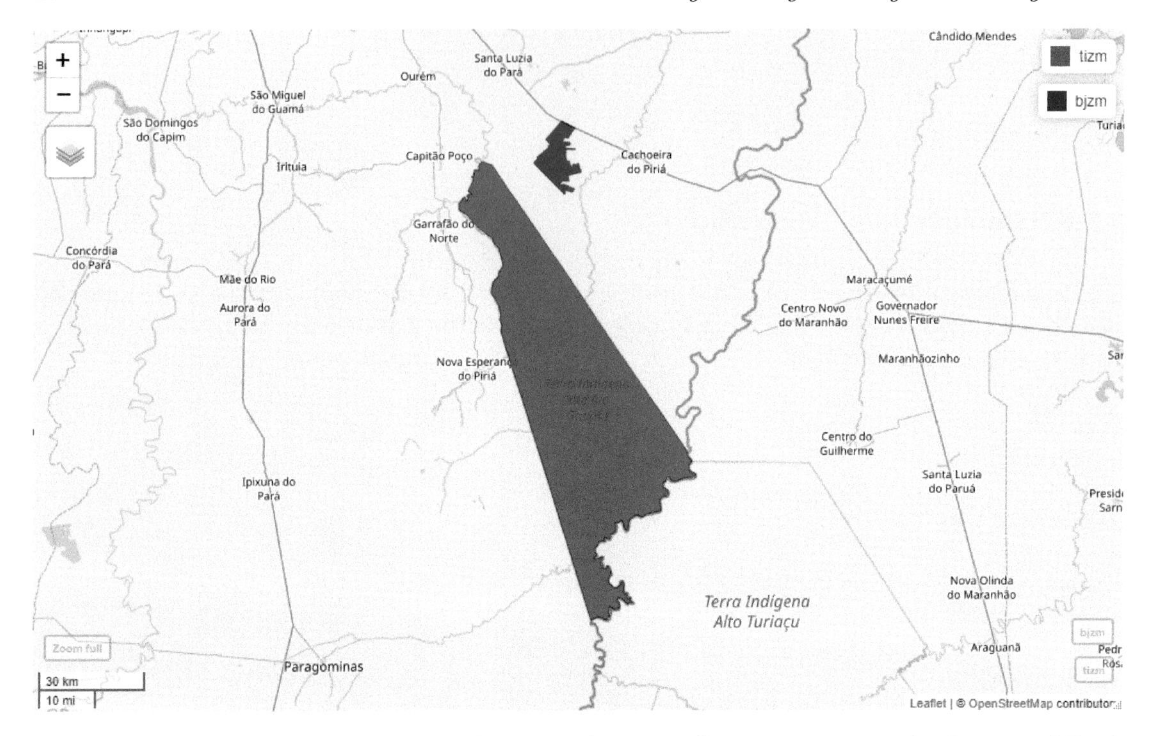

FIGURE 2.1 Mapping of Bom Jesus farm and Alto Rio Guamá indigenous land, state of Pará, Brazil, using R package `mapview`.

The columns `id` and `name` are created and the bracket operator is used to select only the created columns and geometry in a single file named `polsf`. This polygon is used as a reference in later mappings.

```
single_sf$id = c(1:2)
single_sf$name = c("Terra indígena Alto Rio Guamá","Fazenda Bom Jesus")
single_sf
#Simple feature collection with 2 features and 4 fields
#Geometry type: POLYGON
#Dimension:     XY
#Bounding box:  xmin: -47.01409 ymin: -2.883107 xmax: -46.41725 ymax: -1.662143
#Geodetic CRS:  WGS 84
#       Name Description                          geometry id                          name
#1 Trilha 0020           POLYGON ((-46.87984 -1.8406...  1 Terra indígena Alto Rio Guamá
#2                       POLYGON ((-46.75901 -1.6621...  2             Fazenda Bom Jesus
```

Other options can be used to create vectors in R and Earth Engine.

2.1.11 Creating vector data in R and Earth Engine

Data analysis is performed using R packages and cloud computing via the `rgee` package (Aybar et al., 2022).

2.1.11.1 Packages used

The `rgee` (Aybar et al., 2022) and `sf` (Pebesma et al., 2021) packages are required to perform the task and can be loaded for analysis with the `library` function.

```
library(rgee)
library(sf)
```

Access to the data available on the Earth Engine platform is set with the `ee_initialize` function.

```
ee_Initialize(drive = TRUE)
```

2.1.12 Converting polygons to class ee

The polygons used in defining locations of interest in the region are converted into objects of class `ee`. The `sf_as_ee` function is used in converting polygons used in mapping.

```
pols <- sf_as_ee(single_sf)
```

Note that the pipe operator (%>%) can be used to perform the same function as above.

```
pols <- single_sf %>% sf_as_ee()
```

2.1.13 Mapping polygons in R and Earth Engine

Interactive mapping of polygons is performed with `ee` class data. The `Map$setCenter` function is used to set a point to center and zoom into the region of interest. The `Map$addLayer` function is used to map the vector features. A background image from the `Esri.WorldImagery` template is used as a geovisualization element (Figure 2.2).

```
Map$setCenter(-46.68,-2.3, 9)
pol<-Map$addLayer(pols, list(color='#C1CDCD'),'Polígonos')
pol
```

2.1.14 Creating point in R and Earth Engine

Geodetic points and polygons can be created by holding longitude and latitude coordinate values. An `ee.Geometry` file of type point is created in the Jaguari Valley, near the Itaquai River, state of Amazonas, Brazil, where the disappearance of Brazilian indigenist Bruno Pereira and English reporter Dom Phillips was reported.

```
p1 <- ee$Geometry$Point(c(-70.2572273423297, -4.540776134712902))
p1
#EarthEngine Object: Geometry
```

FIGURE 2.2 Mapping polygons of interest in R and Earth Engine.

The object with point geometry is converted to a feature with the `ee$Feature` function. Information about the created point is obtained with the `getInfo()` option.

```
pt1 <- ee$Feature(p1, list('id' = 1, 'name' = 'Vale do Jaguari'))
pt1
#EarthEngine Object: Feature
pt1$getInfo()
#$type
#[1] "Feature"
#$geometry
#$geometry$type
#[1] "Point"
#$geometry$coordinates
#[1]  -3.992078 -70.291167
#$properties
#$properties$id
#[1] 1
#$properties$name
#[1] "Vale do Jaguari"
```

2.1.15 Creating polygon in R and Earth Engine

Similarly, a polygon geometry is created around the point of interest from geodetic coordinates. The `ee$Geometry$Polygon` function is used to create the polygon.

```
pl1 <- ee$Geometry$Polygon(list(c(-70.27139, -4.53142),c(-70.23362, -4.53142),
    c(-70.23362, -4.55384),c(-70.27139, -4.55384),c(-70.27139, -4.53142)))
```

A polygon feature is created from the previously defined geometry with `id` and `name` attributes.

```
pol1 <- ee$Feature(pl1, list('id' = 1, 'name' = 'Pol'))
pol1$getInfo()
#$type
#[1] "Feature"
#$geometry
#$geometry$type
#[1] "Polygon"
#$geometry$coordinates
#$geometry$coordinates[[1]]
#$geometry$coordinates[[1]][[1]]
#[1] -70.27139  -4.55384
#$geometry$coordinates[[1]][[2]]
#[1] -70.23362  -4.55384
#$geometry$coordinates[[1]][[3]]
#[1] -70.23362  -4.53142
#$geometry$coordinates[[1]][[4]]
#[1] -70.27139  -4.53142
#$geometry$coordinates[[1]][[5]]
#[1] -70.27139  -4.55384
#$properties
#$properties$id
#[1] 1
#$properties$name
#[1] "Pol"
```

2.1.16 Mapping point and polygon in R and Earth Engine

The point and polygon are mapped over the region with the `Map$addLayer` function and centered relative to the polygon with the `Map$centerObject` function. A background image from the `Esri.WorldImagery` template is also used for geovisualization. It can be seen that the Itaquai River flows through Atalaia do Norte, bordering Peru and Colombia. The geographical location is strategic and difficult to access, according to the complexity of the terrain composed of forests, rivers and **indigenous**[1] communities. In addition, data and **reports**[2] conducted with the local population denounce historical problems of regional misdemeanors and conflicts with indigenous populations migrating through the region. Given the complexity of the local conflict in the region, Bruno Pereira and Dom Phillips were murdered and quartered in this region in June 2022. As can be seen, the Itaquai River is a river with many meanders, indicating a high level of sinuosity. The natural meanders of a river indicate the age of ancient rivers, the river channel constantly changes position along its stretch on the plain, through a continuous process of erosion and deposition of solids on its banks, mainly sand (Figure 2.3).

[1] https://youtu.be/8z4I0Dondvs
[2] https://youtu.be/bWWIHDGCKoQ

```
Map$centerObject(pol1)
Map$addLayer(pt1,list(color='red'), name = 'Point feature')+
Map$addLayer(pol1,list(color='#C1CDCD'), name = 'Polygon feature')
```

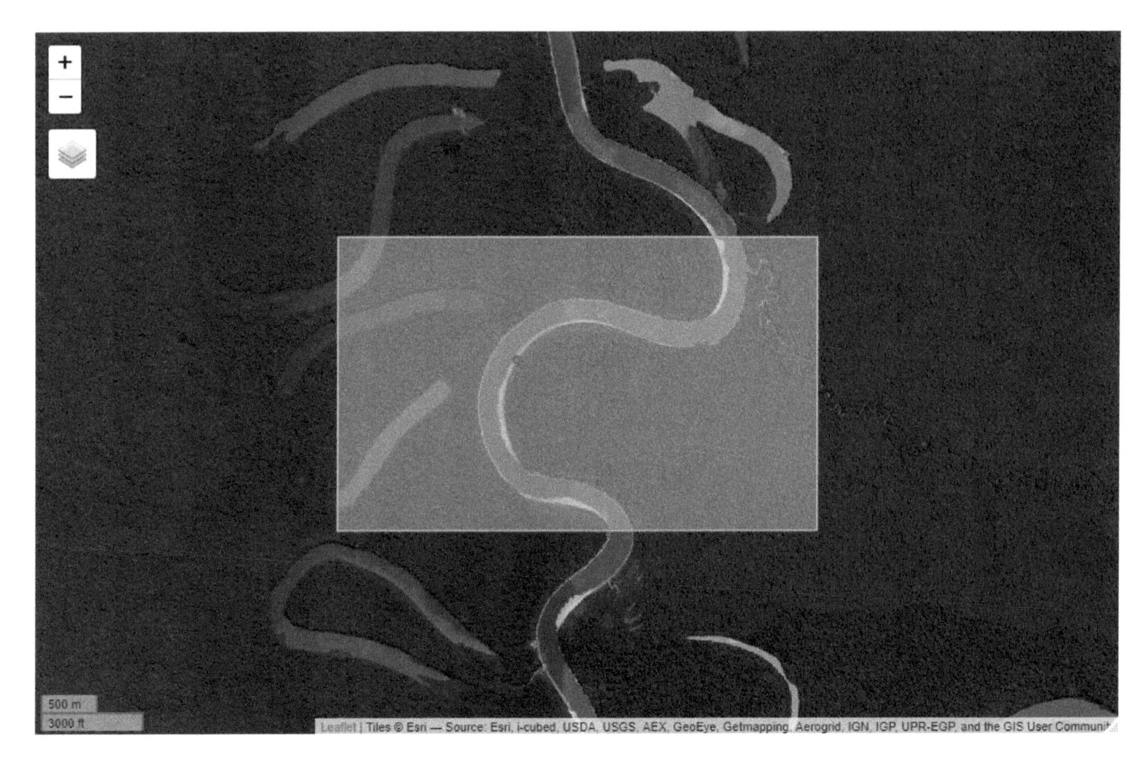

FIGURE 2.3 Point and polygon mapping created in the Jaguari Valley region, state of Amazonas, Brazil, using R and Earth Engine.

2.2 Solved Exercises

2.2.1 Define remote sensing and explain a problem that can be studied with the remote sensing process.

A: Remote sensing is the art, science, and technology of obtaining reliable information about physical objects and environment through the process of recording, measuring, and interpreting images and digital representations of energy patterns derived from non-contact sensor systems. One problem that can be studied in remote sensing is to perform productivity monitoring of coffee fruits. Coffee yield can be simply monitored by spectral agrometeorological model, using meteorological variables obtained by remote sensing of vegetation indices obtained from satellite monitoring with multispectral sensors.

2.2.2 Define radiometric data that can be used in the remote sensing process.

A: Normalized Difference Vegetation Index (NDVI) from the MODIS sensor.

2.2.3 Define a variable that can be obtained *in situ* to compare with radiometric data in the remote sensing process.

A: Volume of harvested coffee berries gauged by a graduated container.

2.2.4 Explain the advantages and disadvantages of remote sensing.

A: Advantages of remote sensing: Aerial perspective on a global, national and regional scale; historical collection of images; perception beyond human vision; extraction of three-dimensional landform information. Limitations of remote sensing: Can be expensive; inaccurate; out of calibration; complex; does not solve all problems.

2.2.5 Name the resolutions used in remote sensing.

A: Spatial, spectral, radiometric, and temporal.

2.2.6 Which of these characteristics favors orbital remote sensing over aerial remote sensing?

a. Spatial resolution.
b. Cost.
c. Time resolution. [X]
d. Stereoscopic view.

2.2.7 Which of these characteristics favors aerial sensing over orbital sensing?

a. Spatial resolution. [X]
b. Spectral resolution.
c. Time resolution.
d. Stereoscopic view.

2.2.8 Cite two packages used to import raster and vector data in R.

A: R packages `raster` and `sf`.

2.2.9 Cite R functions that can be used to stack bands from the OLI sensor of the Landsat-8 satellite?

A: stack(); brick().

2.2.10 What are the basic differences between raster data and vector data in R?

A: A remote sensing raster object refers to an image with number of columns, rows, pixels, extent, spatial resolution, and reference coordinate system. A vector object is discrete with a description of geometry and with the possibility of including additional variables called "attributes".

2.2.11 Create a graph representing the temporal variation of publication of scientific articles with the keyword remote sensing.

The `structure(list)` function is used to create the file structure with time and article publication data in a `data.frame`.

```
A=structure(list(Years=c(1964, 1965, 1966, 1967, 1968, 1969, 1970, 1971, 1972, 1973,
                        1974, 1975, 1976, 1977, 1978, 1979, 1980,1981, 1982, 1983,
                        1984, 1985, 1986, 1987, 1988, 1989, 1990, 1991, 1992, 1993,
                        1994, 1995, 1996, 1997, 1998, 1999, 2000, 2001, 2002, 2003,
                        2004, 2005, 2006, 2007, 2008, 2009, 2010, 2011, 2012, 2013,
                        2014, 2015, 2016, 2017, 2018),
                 Articles = c(7, 4, 13, 21, 31, 39, 130, 116, 86, 153, 232, 232, 269,
                          314, 350, 466, 589, 514, 603, 704, 791, 854, 822, 1069,
                          1040, 1212, 1148, 1256, 1329, 1287, 1324, 1432, 1621,
                          1581, 1087, 1066, 1393, 1525, 1508, 1775, 1769, 2166,
                          2312, 2591, 2797, 2880, 2962, 3648, 3970, 4604, 5039,
                          6018, 6393, 6265, 7639)),
            .Names = c("Years", "Articles"),
            row.names = c(1964L, 1965L, 1966L, 1967L, 1968L, 1969L, 1970L, 1971L,
                      1972L, 1973L, 1974L, 1975L, 1976L, 1977L, 1978L, 1979L,
                      1980L, 1981L, 1982L, 1983L, 1984L, 1985L, 1986L, 1987L,
                      1988L, 1989L, 1990L, 1991L, 1992L, 1993L, 1994L, 1995L,
                      1996L, 1997L, 1998L, 1999L, 2000L, 2001L, 2002L, 2003L,
                      2004L, 2005L, 2006L, 2007L, 2008L, 2009L, 2010L, 2011L,
                      2012L, 2013L, 2014L, 2015L, 2016L, 2017L, 2018L),
            class="data.frame")
```

Column names are used to represent the data. The `attach` function is used so that objects can be accessed in the database by names.

```
attach(A)
```

The names of data columns are obtained with the `names` function.

```
names(A)
```

A mathematical exponential regression model is fitted to the data with the `lm` function.

```
exponential.model <- lm(log(Articles)~ Years)
```

An object is created to store the years to be predicted with the `seq` function.

```
Years <- seq(1964, 2018, 1)
```

An exponential model fit is used to estimate values in a line fitted with the `exp(predict)` function. The `plot` and `lines` functions are drawn on the graph of the number of publications as a function of time.

```
Articles.exponential2=exp(predict(exponential.model,list(Time=Years)))
plot(Years, Articles, pch=16, xlab = "Time (Year)", ylab = "Publications")
lines(Years, Articles.exponential2,lwd=2, col="red", xlab="Years", ylab="Publications")
```

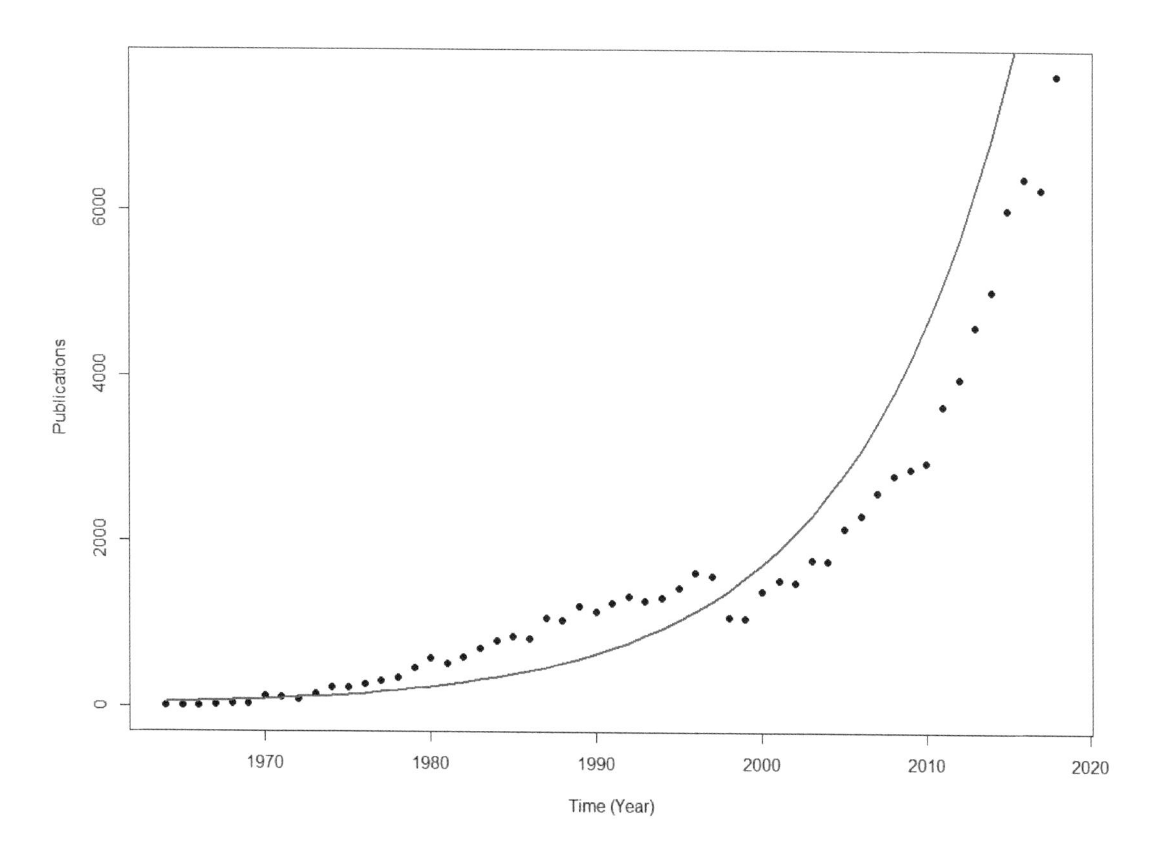

FIGURE 2.4 Temporal variation of publication of scientific articles with the keyword remote sensing.

3

Remote Sensing of Electromagnetic Radiation

3.1 Homework Solution

3.1.1 Subject

Monitoring radiant temperature variation in a geographic region of interest with remote sensing data.

3.1.2 Abstract

Temperature monitoring through remote sensing can be useful for several agricultural, urban and environmental applications. The objective is to use using remote sensing monitoring data in the thermal infrared region to evaluate the surface temperature of the Federal University of Lavras (UFLA) campus. A data processing flow for calibration of radiant temperature data into physical values is performed. The mapping of temperature in Kelvin and degrees Celsius is performed in an urbanized region of the UFLA campus for analysis of heat islands detected with the Landsat-8 thermal band 10, TIR sensor. Places with urban constructions presented higher temperature when compared to vegetation and water regions monitored in the region.

Keywords: Calibration, Landsat-8 TIR, mapping, temperature.

3.1.3 Introduction

Temperature monitoring can be crucial in the development of phenological phases of agricultural crops, and is a critical factor in processes such as flower abortion, choice of crops and varieties best adapted for a region, definition of best sowing period, management and cultivation practices, crop processing activities, and other scientific and socioeconomic applications. The estimation of climatological water balance from monthly mean temperature and rainfall data is an example of how climate information can be indirectly used for socioeconomic purposes, particularly for agriculture, water deficit, potential and actual evapotranspiration, quantifying plant water stress levels, and estimating aridity indices and productivity.

3.1.4 Objective

Obtain Landsat-8 TIR data to calculate the radiant temperature of targets on the Federal University of Lavras campus.

3.1.5 Data acquisition

A practical solution to start remote sensing applications is to use the Landsat satellite to evaluate an object or process in the region chosen for a particular practical work. In this case, to obtain the information in radiance and thermal image, it is necessary to use the "Landsat Collection 1 Level-1" option and download the files instantly. The login page of the Earth Explorer website can be accessed at **login**[1].

3.1.6 Image processing

Landsat-8 TIRS data with metadata are obtained from the image catalog on the Earth Explorer website and used as an example. The radiant temperature calculation is performed from Landsat-8 TIRS thermal band 10.

3.1.7 Packages used

R software is used as an example; however, other programs can be used to accomplish this task at the student's convenience. The packages used are `raster` (Hijmans et al., 2020), `RStoolbox` (Leutner et al., 2019), `rgdal` (Bivand et al., 2021), and `mapview` (Appelhans et al., 2020), and are enabled with the `library` function in the R console.

```
library(raster)
library(RStoolbox)
library(rgdal)
library(mapview)
```

3.1.8 Evaluating the image metadata

The image metadata is also obtained using the `readMeta` function. With this, it was possible to evaluate the image metadata to obtain the parameters for transforming band 10 digital numbers into radiant temperature.

```
metaData <- readMeta(
"C:/geo/artigoML/Landsat/L1C1/LC08_L1TP_218075_20140406_20170424_01_T1_MTL.txt")
summary(metaData)
#Scene:       LC82180752014096LGN01
#Satellite:   LANDSAT8
#Sensor:      OLI_TIRS
#Date:        2014-04-06
#Path/Row:    218/75
#Projection: +proj=utm +zone=23 +datum=WGS84 +units=m +no_defs
#Data:
#                                               FILES QUANTITY CATEGORY
#B1_dn    LC08_L1TP_218075_20140406_20170424_01_T1_B1.TIF       dn    image
#B2_dn    LC08_L1TP_218075_20140406_20170424_01_T1_B2.TIF       dn    image
```

[1] https://ers.cr.usgs.gov/login/

```
#B3_dn    LC08_L1TP_218075_20140406_20170424_01_T1_B3.TIF       dn    image
#B4_dn    LC08_L1TP_218075_20140406_20170424_01_T1_B4.TIF       dn    image
#B5_dn    LC08_L1TP_218075_20140406_20170424_01_T1_B5.TIF       dn    image
#B6_dn    LC08_L1TP_218075_20140406_20170424_01_T1_B6.TIF       dn    image
#B7_dn    LC08_L1TP_218075_20140406_20170424_01_T1_B7.TIF       dn    image
#B9_dn    LC08_L1TP_218075_20140406_20170424_01_T1_B9.TIF       dn    image
#B10_dn LC08_L1TP_218075_20140406_20170424_01_T1_B10.TIF        dn    image
#B11_dn LC08_L1TP_218075_20140406_20170424_01_T1_B11.TIF        dn    image
#B8_dn    LC08_L1TP_218075_20140406_20170424_01_T1_B8.TIF       dn      pan
#QA_dn  LC08_L1TP_218075_20140406_20170424_01_T1_BQA.TIF        dn       qa
#Available calibration parameters (gain and offset):
#   dn -> radiance (toa)
#   dn -> reflectance (toa)
#   dn -> brightness temperature (toa)
```

With this, it is possible to see, among other information, the date and map projection of the Landsat-8 TIRS imaging and image acquisition, on April 06, 2014, "+proj=utm +zone=23 +datum=WGS84 +units=m +no_defs", respectively.

With the `metaData` object created, it is possible to observe more detailed information about the data.

```
metaData
#$METADATA_FILE
#[1] "C:/geo/artigoML/Landsat/L1C1/LC08_L1TP_218075_20140406_20170424_01_T1_MTL.txt"
#$METADATA_FORMAT
#[1] "MTL"
#$SATELLITE
#[1] "LANDSAT8"
#$SENSOR
#[1] "OLI_TIRS"
#$SCENE_ID
#[1] "LC82180752014096LGN01"
#$ACQUISITION_DATE
#[1] "2014-04-06 12:57:55 GMT"
#$PROCESSING_DATE
#[1] "2017-04-24 GMT"
#$PATH_ROW
#path  row
# 218   75
#$PROJECTION
#Coordinate Reference System:
#Deprecated Proj.4 representation:
# +proj=utm +zone=23 +datum=WGS84 +units=m +no_defs
#WKT2 2019 representation:
#PROJCRS["unknown",
#    BASEGEOGCRS["unknown",
#        DATUM["World Geodetic System 1984",
#            ELLIPSOID["WGS 84",6378137,298.257223563,
#                LENGTHUNIT["metre",1]],
#            ID["EPSG",6326]],
```

```
#         PRIMEM["Greenwich",0,
#             ANGLEUNIT["degree",0.0174532925199433],
#             ID["EPSG",8901]]],
#     CONVERSION["UTM zone 23N",
#         METHOD["Transverse Mercator",
#             ID["EPSG",9807]],
#         PARAMETER["Latitude of natural origin",0,
#             ANGLEUNIT["degree",0.0174532925199433],
#             ID["EPSG",8801]],
#         PARAMETER["Longitude of natural origin",-45,
#             ANGLEUNIT["degree",0.0174532925199433],
#             ID["EPSG",8802]],
#         PARAMETER["Scale factor at natural origin",0.9996,
#             SCALEUNIT["unity",1],
#             ID["EPSG",8805]],
#         PARAMETER["False easting",500000,
#             LENGTHUNIT["metre",1],
#             ID["EPSG",8806]],
#         PARAMETER["False northing",0,
#             LENGTHUNIT["metre",1],
#             ID["EPSG",8807]],
#         ID["EPSG",16023]],
#     CS[Cartesian,2],
#         AXIS["(E)",east,
#             ORDER[1],
#             LENGTHUNIT["metre",1,
#                 ID["EPSG",9001]]],
#         AXIS["(N)",north,
#             ORDER[2],
#             LENGTHUNIT["metre",1,
#                 ID["EPSG",9001]]]]
#$SOLAR_PARAMETERS
#  azimuth elevation  distance
#50.308917 48.726904  1.000737
#$DATA
#  FILES  BANDS QUANTITY CATEGORY NA_VALUE SATURATE_VALUE SCALE_FACTOR
#B1_dn    LC08_L1TP_218075_20140406_20170424_01_T1_B1.TIF  B1_dn dn  image NA  NA  1
#B2_dn    LC08_L1TP_218075_20140406_20170424_01_T1_B2.TIF  B2_dn dn  image NA  NA  1
#B3_dn    LC08_L1TP_218075_20140406_20170424_01_T1_B3.TIF  B3_dn dn  image NA  NA  1
#B4_dn    LC08_L1TP_218075_20140406_20170424_01_T1_B4.TIF  B4_dn dn  image NA  NA  1
#B5_dn    LC08_L1TP_218075_20140406_20170424_01_T1_B5.TIF  B5_dn dn  image NA  NA  1
#B6_dn    LC08_L1TP_218075_20140406_20170424_01_T1_B6.TIF  B6_dn dn  image NA  NA  1
#B7_dn    LC08_L1TP_218075_20140406_20170424_01_T1_B7.TIF  B7_dn dn  image NA  NA  1
#B9_dn    LC08_L1TP_218075_20140406_20170424_01_T1_B9.TIF  B9_dn dn  image NA  NA  1
#B10_dn LC08_L1TP_218075_20140406_20170424_01_T1_B10.TIF B10_dn dn  image NA  NA  1
#B11_dn LC08_L1TP_218075_20140406_20170424_01_T1_B11.TIF B11_dn dn  image NA  NA  1
#B8_dn    LC08_L1TP_218075_20140406_20170424_01_T1_B8.TIF  B8_dn dn    pan NA  NA  1
#QA_dn  LC08_L1TP_218075_20140406_20170424_01_T1_BQA.TIF  QA_dn dn     qa NA  NA  1
#        DATA_TYPE SPATIAL_RESOLUTION RADIOMETRIC_RESOLUTION
#B1_dn          NA                 30                     16
#B2_dn          NA                 30                     16
#B3_dn          NA                 30                     16
```

```
#B4_dn          NA                  30                      16
#B5_dn          NA                  30                      16
#B6_dn          NA                  30                      16
#B7_dn          NA                  30                      16
#B9_dn          NA                  30                      16
#B10_dn         NA                  30                      16
#B11_dn         NA                  30                      16
#B8_dn          NA                  15                      16
#QA_dn          NA                  30                      16
#$CALRAD
#           offset       gain
#B1_dn   -62.68630 0.01253700
#B2_dn   -64.19154 0.01283800
#B3_dn   -59.15195 0.01183000
#B4_dn   -49.88026 0.00997610
#B5_dn   -30.52423 0.00610480
#B6_dn    -7.59110 0.00151820
#B7_dn    -2.55861 0.00051172
#B8_dn   -56.45072 0.01129000
#B9_dn   -11.92956 0.00238590
#B10_dn    0.10000 0.00033420
#B11_dn    0.10000 0.00033420
#$CALREF
#       offset  gain
#B1_dn    -0.1 2e-05
#B2_dn    -0.1 2e-05
#B3_dn    -0.1 2e-05
#B4_dn    -0.1 2e-05
#B5_dn    -0.1 2e-05
#B6_dn    -0.1 2e-05
#B7_dn    -0.1 2e-05
#B8_dn    -0.1 2e-05
#B9_dn    -0.1 2e-05
#$CALBT
#              K1       K2
#B10_dn 774.8853 1321.079
#B11_dn 480.8883 1201.144
#attr(,"class")
#[1] "ImageMetaData" "RStoolbox"
```

3.1.9 Recording gain and offset values of the thermal band

The gain and offset values of the thermal band number 10 radiance and the K_1 and K_2 calibration parameters of the radiometric temperature are recorded for applying mathematical models.

```
RADIANCE_MULT_BAND_10 <- 3.3420E-04
RADIANCE_ADD_BAND_10 <- 0.10000
K1_CONSTANT_BAND_10 <- 774.8853
K2_CONSTANT_BAND_10 <- 1321.079
```

3.1.10 Importing band 10 into R

Band 10 is imported into R with the `raster` function.

```
band_10 <- raster(
"C:/geo/artigoML/Landsat/L1C1/LC08_L1TP_218075_20140406_20170424_01_T1_B10.TIF")
```

3.1.11 Converting band 10 into radiance at the top of the atmosphere

A function is created to convert the band 10 digital numbers into radiance at the top of the atmosphere using the `calc` function.

```
toa_band10 <- calc(
band_10,fun=function(x){RADIANCE_MULT_BAND_10 * x + RADIANCE_ADD_BAND_10})
```

3.1.12 Converting the radiance at the top of the atmosphere to brightness temperature

The radiance at the top of the atmosphere is converted to brightness (radiant) temperature at the sensor by applying a mathematical model. Then another function was used to convert the temperature into degrees Celsius.

```
# Convert temperature to Kelvin
temp10_kelvin <- calc(
toa_band10, fun=function(x){K2_CONSTANT_BAND_10/log(K1_CONSTANT_BAND_10/x + 1)})
# Convert Kelvin temperature to degrees Celsius
temp10_celsius <- calc(temp10_kelvin, fun=function(x){x - 273.15})
```

3.1.13 Reprojecting the image

The temperature image in Celsius is reprojected to the Southern Hemisphere using a parameter of the `crs` argument in UTM projection, WGS-84 ellipsoid, zone 23S. For this, the digital elevation model raster was used as a reference for matching the rasters. This function can consume computational memory for processing, which requires time to complete the whole process.

3.1.14 Defining a subset

The `extent` function is used to define a rectangular geographic region used as a reference to map the temperature in an area near the Federal University of Lavras campus. The `crop` function is used to crop the temperature image in the region of interest.

```
e<-extent(501789.4, 503332, 7652923, 7654290)
temp<-crop(tempprj, e)
```

3.1.15 Exporting the temperature raster layer

The Landsat-8 OLI radiometric temperature data, in degrees Celsius, in the area of interest are exported in `geotiff` format.

```
writeRaster(temp, filename="C:/sr/c2/temp.tif",
            format="GTiff", overwrite=TRUE)
```

3.1.16 Importing the temperature raster layer

The image with temperature variation is imported into R with the `raster` function and renamed with the `names` function.

```
temp <- raster("C:/sr/c2/temp.tif")
names(temp) <- c('temp')
```

3.1.17 Mapping the results

The temperature image in a region near the Federal University of Lavras campus is mapped with the `spplot` function (Figure 3.1).

```
spplot(temp, "temp", scales = list(draw = T))
```

The results can be evaluated in a comparative way with a very high resolution color composition available in the `mapview` data collection. With this, it is possible to observe heat islands in urban construction sites when compared to the tree-covered vegetation and water region contained in this subset of the analyzed Landsat-8 data (Figure 3.2).

```
mapview(temp)
```

3.2 Solved Exercises

3.2.1 Determine the atmospheric transmittance coefficient for two consecutive days of a location at latitude -21.248488°, longitude -45.001375°, and altitude 964 m above sea level.

```
install.packages("solrad") # Install solrad package
```

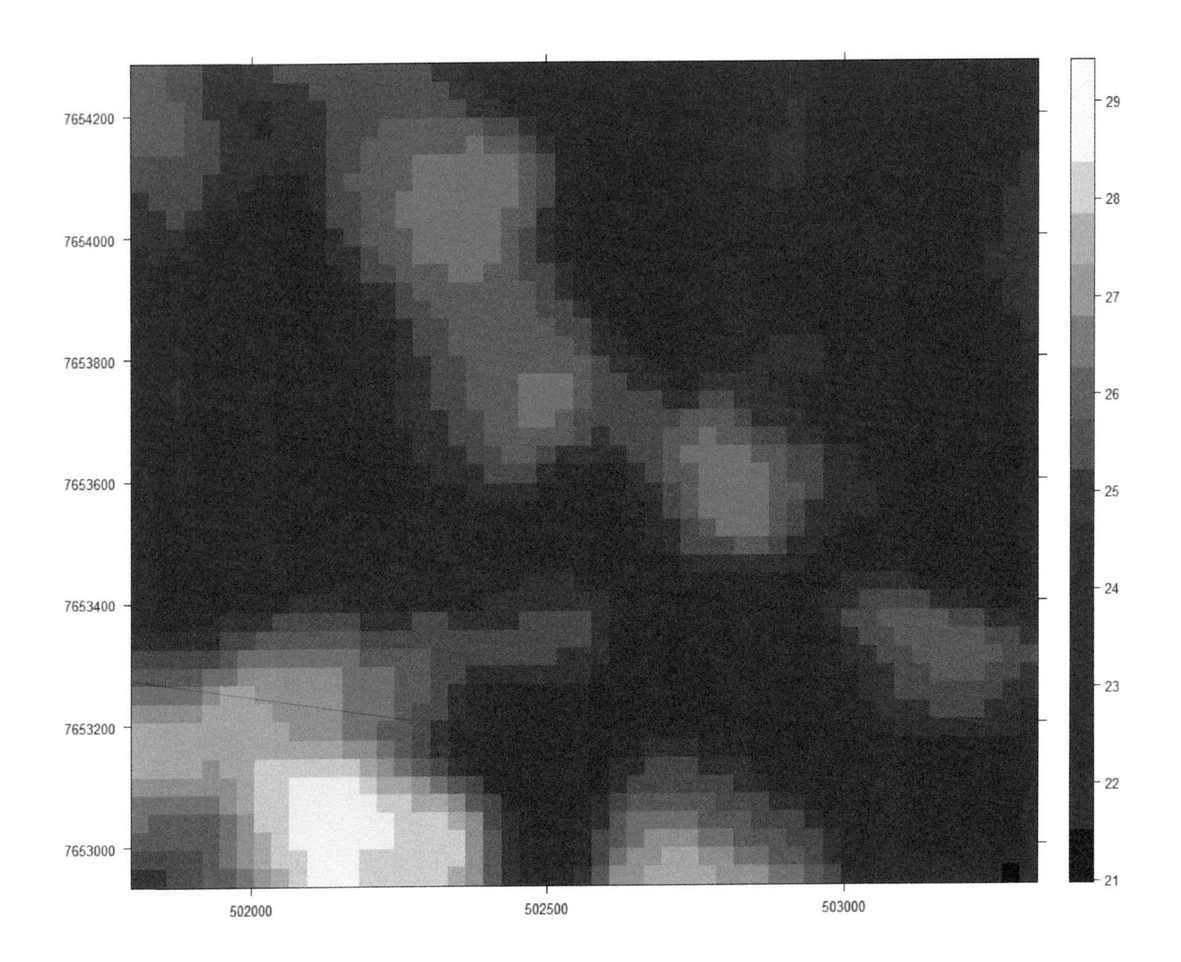

FIGURE 3.1 Temperature mapping from Landsat-8 monitoring on April 6, 2014 in a region near the Federal University of Lavras, state of Minas Gerais, Brazil.

```
library(solrad) # Enable the package
# Define the sequence of days
DOY <- seq(0, 2, 0.05)
# Calculate the atmospheric transmittance coefficient
tb <- Transmittance(DOY, Lat = -21.248488, Lon=-45.001375, SLon=-45.001375, DS=0,
                    Elevation = 964.0)
# Plot the variation of the transmittance coefficient over the days
plot(DOY, tb)
```

The transmittance values ranged from 0 to 0.7 for the first two days of the year (Figure 3.3). More details on the calculation of factors involved with solar extraterrestrial radiation, atmospheric transmittance, solar constant and other solar calculation variables are described in Seyednasrollah et al. (2013).

FIGURE 3.2 Temperature mapping and comparison with a natural color composition from `mapview` package data collection.

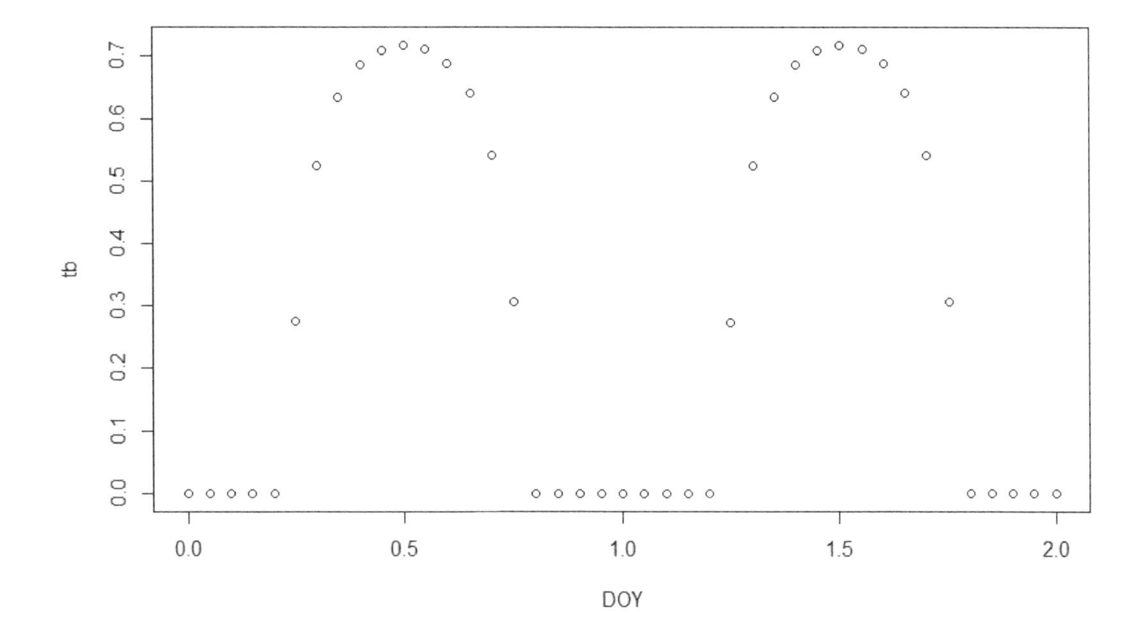

FIGURE 3.3 Coefficient of atmospheric transmittance for two days of the year.

3.2.2 Make a graph of the electromagnetic spectrum.

```
install.packages("SpecHelpers") # Install the SpecHelpers package
```

```
library(SpecHelpers) # Enable the package
# Making a graph of the electromagnetic spectrum with applications
emSpectrum(molecular = TRUE, applications = TRUE)
```

The electromagnetic spectrum graph with applications is created with the `emSpectrum` function of the `SpecHelpers` package (Hanson, 2017) (Figure 3.4).

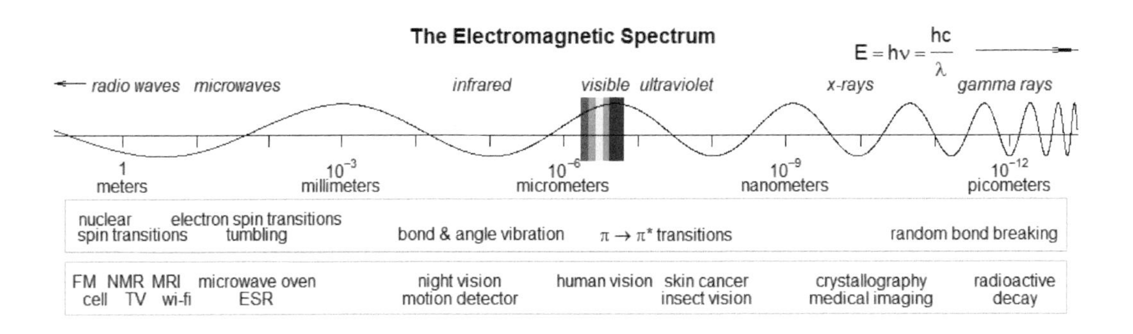

FIGURE 3.4 Electromagnetic spectrum with applications.

3.2.3 Arrange the following wavelengths in ascending order: near-infrared, blue, ultraviolet, radio waves, and X-rays.

A: X-rays, ultraviolet, blue, near-infrared, and radio waves.

3.2.4 An X-ray machine has produced radiation at a wavelength of 1.0 * 10^{-11} m. Determine the frequency of the radiation and the energy. Consider h = 6.63 * 10^{-34} Js and c = 3.0 * 10^{8} ms^{-1}.

A: The frequency is $3e^{+19}$ Hz. The energy is $1.989e^{-14}$ J.

```
#c = lambda*f
f <- (3*10^8)/(1*10^-11)
f
```

```
## [1] 3e+19
```

```
#E = hf
E <- (6.63*10^-34)*(3*10^19)
E
```

```
## [1] 1.989e-14
```

3.2.5 A radio transmitter produced waves with a frequency of 1.0 * 10^8 Hz. Determine the wavelength of the signal in meters. Consider c = 3.0 * 10^8 ms^{-1}.

A: The wavelength was 3 m.

```
#c = lambda*f
lambda <- (3*10^8)/(1*10^8)
lambda
```

```
## [1] 3
```

3.2.6 Determine the solar azimuth and zenith angles in radians from the angular values in decimal degrees of 61.96° and 40.25°, respectively.

A: The angle values in radians of solar azimuth and zenith are 1.081406 and 0.702495, respectively.

```
install.packages("circular") # Install package
```

```
library(circular) # Enable the package
```

```
##
## Attaching package: 'circular'

## The following object is masked _by_ '.GlobalEnv':
##
##     wind

## The following objects are masked from 'package:stats':
##
##     sd, var
```

```
az<-rad(61.96) # solar azimuth
az
```

```
## [1] 1.081406
```

```
z<-90-40.25 # Solar zenith
zs<-rad(z)
zs
```

```
## [1] 0.8683013
```

3.2.7 Why is it important to know about emissivity when conducting a remote sensing survey in the thermal infrared?

A: Emissivity is important because two objects very close to each other on the ground can have the same kinetic (true) temperature, but different apparent temperatures when measured by a thermal radiometer, just by having different emissivities.

3.2.8 Why is knowing the dominant wavelength of an object important to remote sensing in the thermal infrared?

A: The dominant wavelength provides important information about the part of the electromagnetic spectrum that should be chosen to remotely observe an object. For example, forest fires at 800 K have a dominant wavelength of 3.62 μm, so the most appropriate remote sensing system might be a 3-5 μm thermal infrared detector. In the case of soil, water and rock monitoring at the ambient temperature of the Earth's surface (300 K) and dominant wavelength of 9.67 μm, it would be more appropriate to use a thermal infrared detector in the region of 8-14 μm.

3.2.9 Define hemispheric reflectance and explain how to use this variable in the remote sensing process.

A: Reflectance is the ratio of the reflected energy flux to the total energy flux incident on the remotely sensed target. By using reflectance data of an object at different wavelengths it is possible to realize the spectral signature of the object, which can be useful in establishing criteria for comparing the target against an existing standard.

3.2.10 Define path radiance and explain its effect on remote sensing data and image interpretation.

A: Path radiance is a type of radiation that is not desired in the remote sensing process, because this type of radiance has suffered from atmospheric interference and can determine the formation of images with poorer quality of interpretation.

3.2.11 Which of the following variables represents the energy detected by a sensor system on a satellite?

a. Irradiance.
b. Radiance. [X]
c. Absorptance.
d. Transmittance.

3.2.12 Reflectance is defined as the ratio between:

a. Reflected and emitted flux.
b. Reflected and backscattered flux.
c. Reflected and incident flow. [X]
d. Reflected and absorbed flux.

3.2.13 What are objects with similar radiance from different observation angles called?

a. Backscattered.
b. Lambertians. [X]
c. Specular.
d. Amorphous.

3.2.14 Determine the dominant wavelength of the Sun and Earth (micrometers) considering that these objects approach black bodies at temperatures of 6000 and 300 K, respectively.

```
# Lambda max = k/T
lambdaMaxSun <- 2898/6000 # Sun
lambdaMaxSun
```

```
## [1] 0.483
```

```
lambdaMaxEarth <- 2898/300 # Earth
lambdaMaxEarth
```

```
## [1] 9.66
```

3.2.15 Determine the total emissive energy power of the Sun and Earth (W m^{-2}) considering that these objects approach black bodies at temperatures of 6000 and 300 K, respectively.

```
# Mlambda <- sigma*T^4
MlambdaSun <- (5.6697e-8)*(6000^4)
MlambdaSun
```

```
## [1] 73479312
```

```
MlambdaEarth <- (5.6697e-8)*(300^4)
MlambdaEarth
```

```
## [1] 459.2457
```

4

Remote Sensing Sensors and Satellite Systems

4.1 Homework Solution

4.1.1 Subject

Monitoring the Earth's surface with free, very high spatial resolution remote sensing data.

4.1.2 Abstract

CBERS-04A, the sixth satellite in the CBERS family, was successfully launched and placed into orbit on December 20, 2019, by the Long March 4B rocket from the Taiyuan Satellite Launch Center in China. CBERS imaging data can be used in diverse applications for monitoring rural and urban areas, soil, rocks, water, and vegetation on the surface. The WPM camera is the main payload of the Chinese-made CBERS-04A, and aims to provide images with 2-m panoramic resolution and 8-m multispectral resolution simultaneously in the satellite's orbit. CBERS-04A monitoring data from the WPM camera is used as an example of panchromatic band mapping at 2-m spatial resolution in monitoring conducted on July 8, 2020 at the Funil dam. By processing the images and mapping the results, it is possible to visualize details of urban buildings and natural targets in the region, such as vegetation and water.

Keywords: CBERS-04A, image processing, mapping, panchromatic band.

4.1.3 Introduction

Surface monitoring by Earth observation satellites can be crucial for investigating details about agricultural, urban targets, and conducting technical forensics in some situations. Remote sensing and image processing applications have been performed with data from the CBERS-04A mission for wildfire detection in the Pantanal in the state of Mato Grosso (Higa et al., 2022), vegetation mapping (Marinho et al., 2022), and soil and water conservation studies (Goes & Oliveira, 2022).

4.1.4 Objective

The objective of the homework is to obtain remote sensing data and perform a preliminary mapping of this data at a location of interest with image processing software.

4.1.5 Data acquisition

CBERS-04A monitoring data from the WPM camera (CBERS4A_WPM_L4_DN) are used as an example of panchromatic band mapping at 2-m spatial resolution in monitoring performed on July 8, 2020 at the Funil dam. The data obtained refer to the L4 processing level, referring to an orthorectified image with radiometric and geometric correction using control points and a digital elevation model. More details on how to obtain the data can be obtained in a **data catalog**[1] on the Internet.

In this case, the software R is used as an example, however other programs can be used to accomplish this task at the student's discretion.

4.1.6 Enabling packages

The `raster` package is enabled for use with the `library` function.

```
library(raster)
```

4.1.7 Importing the panchromatic band

The CBERS-04A panchromatic band is imported into R with the `raster` function from the directory in which the data was stored on the computer.

```
b0a <- raster(
"C:/sr/cbers4A/CBERS_4A_WPM_20200708_201_140_L4/
CBERS_4A_WPM_20200708_201_140_L4_BAND0.tif")
```

4.1.8 Defining a geographic extent for mapping

The `extent` function is used to define a geographic region of interest for mapping.

```
e1<-extent(508650.5, 510148.2, 7660994, 7662325)
```

4.1.9 Mapping the data in R

The `plot` function is used for panchromatic band mapping. The `ext` parameter is used to specify the region of interest to be mapped. A color palette in digital gray levels is specified in the `col` parameter, with the `grey.colors` color palette argument. The `legend` parameter is set with the `FALSE` argument to hide the legend. A title for the image is set in the `main` parameter (Figure 4.1).

```
plot(b0a, ext=e1, col = grey.colors(100, start=0, end=1),
     legend=FALSE, main = "Ijaci, Macaia, MG")
```

[1] http://www2.dgi.inpe.br/catalogo/explore

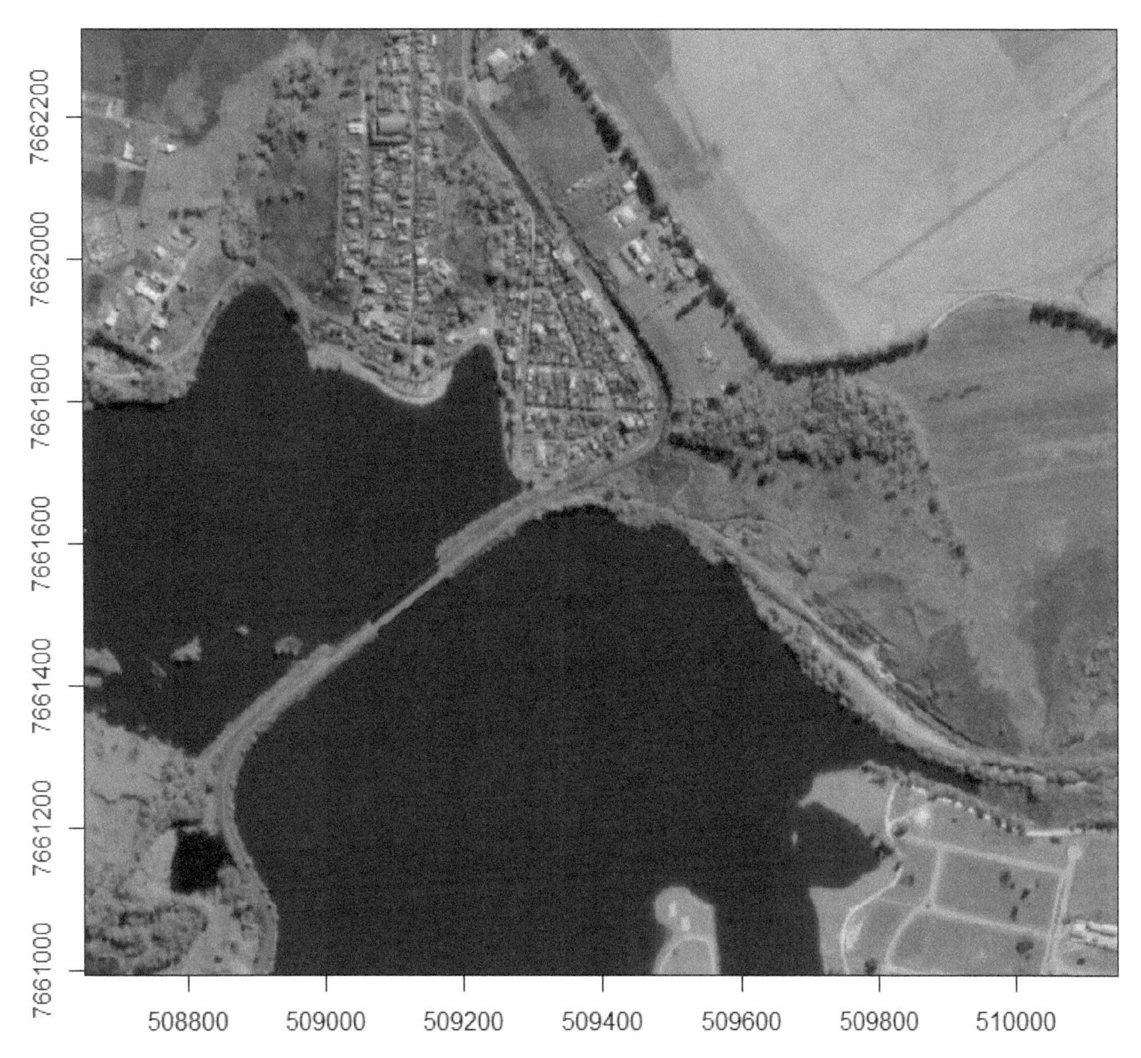

FIGURE 4.1 CBERS-04A panchromatic band mapping in the region of the municipalities of Ijaci and Macaia, state of Minas Gerais, on July 8, 2020.

4.2 Solved Exercises

4.2.1 **Landsat-4 has an inclination angle of 98.3° and an altitude of 687 km. The satellite passes the Equator every 98.5 minutes. Determine the number of revolutions per day (24 h) (orbits) of the satellite around the Earth.**

```
# NRS=T/TE
NRS<-1440/98.5
NRS
```

```
## [1] 14.61929
```

4.2.2 Determine the distance between polar orbits performed by a satellite considering the Earth's circumference at the Equator is approximately 40000 km and that the platform can complete 14.54 orbits per day.

```
# DT = CT/NRS
DT<-40000/14.54
DT
```

```
## [1] 2751.032
```

4.2.3 A coffee crop is mapped using a radiometer with an IFOV of 2.5 mrad at an altitude of 705 km. Determine the spatial resolution of a pixel in the image in meters.

```
# D=2Htan(B/2)
D <- 2*705000*(tan(0.0025/2*pi/180))
D
```

```
## [1] 30.76143
```

4.2.4 Mapping of rural properties was performed with a radiometer with a full angle of view along a 15° scan line at the altitude of 705 km. Determine the imaging swath width on the terrain in meters.

```
# SW=2Htan(theta/2)
SW <- 2*705000*(tan(15/2*pi/180))
SW
```

```
## [1] 185630
```

4.2.5 A soybean crop was imaged using a sensor with a radiometric resolution of 8 bits. Determine how many digital numbers can be registered by the sensor system.

```
# N=2^n
N <- 2^8
N
```

```
## [1] 256
```

4.2.6 How is the radiometric resolution of a sensor measured?

a. Number of spectral bands.
b. Pixel size.
c. Number of bits per pixel. [X]
d. Orbital cycle.

4.2.7 What is the main advantage of a geostationary satellite?

a. Temporal resolution. [X]
b. Spatial resolution.
c. Angular resolution.
d. Spectral resolution.

4.2.8 Which of the following variables represents the energy detected by a sensor system on a satellite?

a. Irradiance.
b. Radiance. [X]
c. Absorptance.
d. Transmittance.

4.2.9 Which of the following resolutions is the most relevant for urban studies?

a. Spatial. [X]
b. Spectral.
c. Temporal.
d. Radiometric.

4.2.10 Define a problem in remote sensing and a sensor to monitor that problem. Justify the choice.

A: Problem: Monitor the vegetative vigor of coffee crops after the coffee harvest; Sensor: Landsat-8 OLI/TIRS; Justification: Landsat-8 images are free and images of a coffee crop are available every 16 days. Through the spatial resolution of 30 m in the reflective spectrum bands and the thermal band of 100 m in the thermal region, it is possible to monitor a large extension of crops in view of the size of the Landsat-8 scene of approximately 180 km^2 and make inferences in the areas with lower vegetation index and higher temperature, humidity and vigor that may have suffered damage during the coffee harvest.

5

Remote Sensing of Vegetation

5.1 Homework Solution

5.1.1 Subject

Determination of vegetation index from multispectral remote sensing data.

5.1.2 Abstract

CBERS-04A has great potential for land surface monitoring in different applications in rural and urban areas. With the Wide Scan Multispectral and Panchromatic (WPM) camera of the CBERS-04A satellite it is possible to obtain panchromatic images at 2-m spatial resolution and multispectral images at 8-m spatial resolution with revisiting the same region every 31 days. A simplified approach is used to determine the normalized difference vegetation index (NDVI) near the Funil dam, Minas Gerais. By processing the images and mapping the results it is possible to visualize the variation of regions with low and high values of vegetation index at the monitored site. Methodologies for pre-processing the data to improve the results obtained by performing digital number conversions for radiance, reflectance at the top of the atmosphere, and surface reflectance should be carried out in future studies.

Keywords: CBERS-04A, mapping, NDVI, R software, `raster` package.

5.1.3 Introduction

Vegetation is a renewable natural resource of landscape significance and economic expression and is part of most terrestrial ecosystems. Earth surface monitoring by Earth observation satellites can be crucial, and multispectral imaging cameras can be used in applications to determine indices that reflect conditions of vegetation vigor. Using remote sensing, it is possible to estimate and evaluate the vegetation cover of both urban and rural areas, as well as provide a record of qualitative or quantitative changes in vegetation over time (Köhler, 1998).

5.1.4 Objective

The objective of the homework is to obtain remote sensing data and perform normalized difference vegetation index determination on image data from the CBERS-04A satellite, WPM camera.

5.1.5 CBERS-04A data specifications

Information about CBERS-04A data characteristics and specifications can be obtained from Chapter 4.

5.1.6 Getting CBERS-04A data

CBERS-04A monitoring data from the WPM camera is used as an example of NDVI vegetation index mapping from monitoring conducted on July 8, 2020 at the Funil dam, state of Minas Gerais. The data used are processed at level 4 (L4), by obtaining an orthorectified image with radiometric correction and geometric correction using control points and a digital elevation model. More details on how to obtain the data can be found in Chapter 4.

The software R is used as an example, but other programs can be used to accomplish this task according to the student's discretion.

5.1.7 Enabling packages

The R package `raster` (Hijmans et al., 2020) is used in this computational practice and it is enabled for use with the `library` function.

```
library(raster)
```

5.1.8 Importing multispectral bands

The bands in the red and near-infrared spectral region of the CBERS-04A are imported into R with the `raster` function from the folder in which the data has been stored on your computer.

```
bred <- raster(
"C:/sr/cbers4A/CBERS_4A_WPM_20200708_201_140_L4_BAND3.tif")
bnir <- raster(
"C:/sr/cbers4A/CBERS_4A_WPM_20200708_201_140_L4_BAND4.tif")
```

5.1.9 Determine the NDVI

The `overlay` function is used together with a function applied in the `fun` argument to calculate the NDVI from a normalized difference equation in R from the red (bred) and NIR (bnir) bands of the CBERS-04A.

```
ndvi <- overlay(bred, bnir, fun = function(x, y) {
  (y-x) / (y+x)
})
```

5.1.10 Define a geographic extent for mapping

The `extent` function is used to define a geographic region of interest for mapping.

```
e1<-extent(508650.5, 510148.2, 7660994, 7662325)
```

5.1.11 Mapping the data in R

The `plot` function is used for mapping the NDVI. The `ext` argument is used to specify the region of interest to be mapped. A terrain color palette is specified in the `col` argument, with the `terrain.colors` color palette. The `legend` argument is set with `TRUE` to show the legend. A title for the image is set in the `main` argument.

```
plot(ndvi, ext=e1, col = rev(terrain.colors(100)),
     legend=TRUE, main = "NDVI, Ijaci, Macaia, MG")
```

Thereby, it is possible to see that urbanized regions with water present low NDVI values, while preservation areas with forest vegetation and central pivot areas with emerald grass cultivation and undergrowth present high NDVI values around 0.6 (Figure 5.1).

5.2 Solved Exercises

5.2.1 Explain how the interaction between visible and infrared radiation occurs in a dicotyledon leaf.

A: The leaf pigments in the palisade parenchyma cells impact the absorption and reflectance of visible (blue, green, and red) light, while in the mesophyll cells they impact the absorption and reflectance of incident infrared energy.

5.2.2 Explain how to calculate the NDVI of vegetation with the Landsat-8 OLI sensor.

A: The normalized difference vegetation index (NDVI) of vegetation at the surface is calculated by:

$$NDVI = \frac{\rho_{NIR} - \rho_{red}}{\rho_{NIR} + \rho_{red}} \tag{5.1}$$

where ρ_{NIR} and ρ_{red} are surface reflectance values obtained in the near-infrared and red bands, respectively.

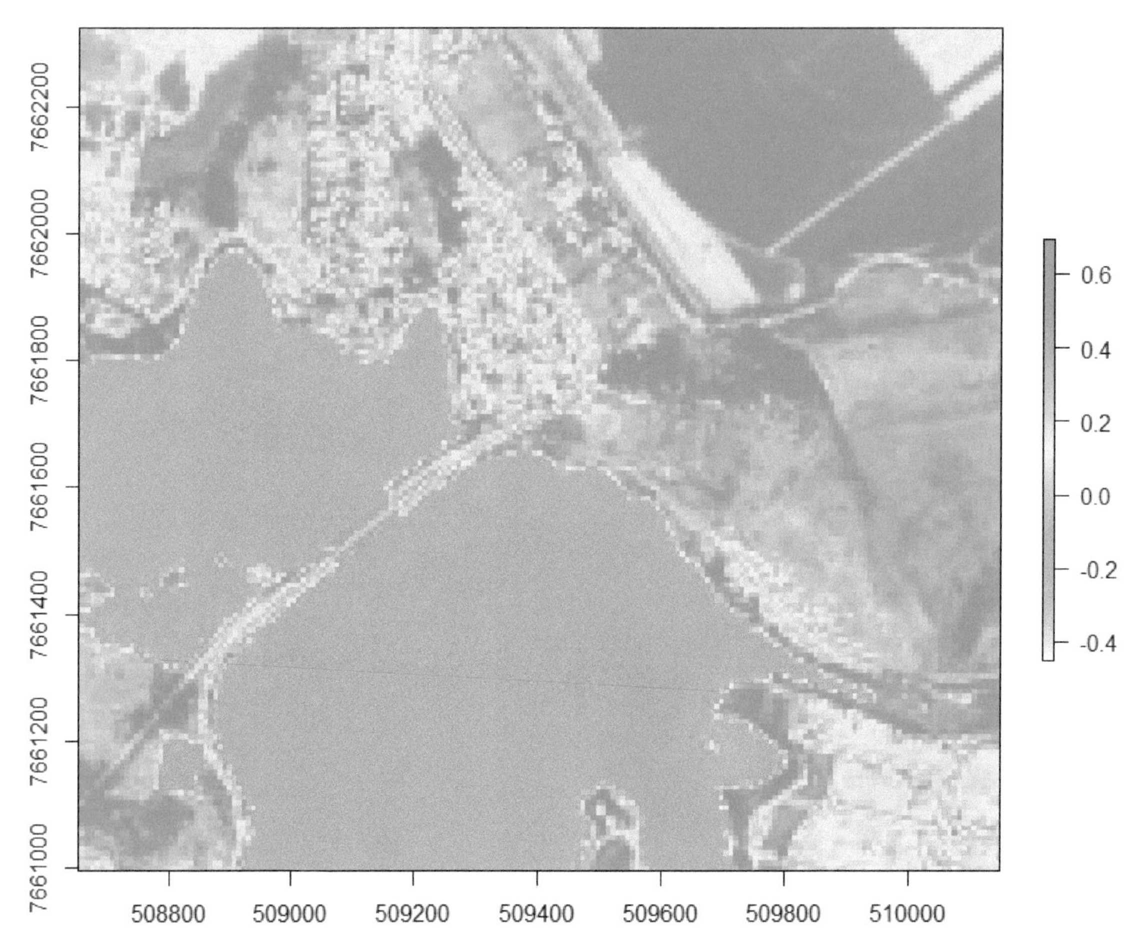

FIGURE 5.1 NDVI mapping of CBERS-04A in the region of the municipalities Ijaci and Macaia, state of Minas Gerais, Brazil, on July 8, 2020.

5.2.3 Cite two factors that can interfere with the spectral response of plants in agricultural crops monitored by multispectral remote sensing.

A: Shade and clouds.

5.2.4 Justify the monitoring of a vegetation type with an index and how this will be done over time.

A: The index vegetation type of a 30-ha coffee crop in the month of August will be monitored with NDVI from Landsat-8 OLI, considering that in this month there is less cloud effect on the reflectance values of the images. Considering that crop defoliation after harvest can affect productivity in the following season, a differentiated crop management will be proposed in areas with higher defoliation and lower NDVI. Monitoring of NDVI in the field will be conducted for a minimum of 2 consecutive years.

5.2.5 Define a remote sensing of vegetation problem and a sensor used to monitor this problem. Justify the choice.

A: Problem: Satellite monitoring of the deforestation of the Brazilian Amazon forest. In the **PRODES**[1] project, satellite monitoring of clearcut deforestation in the Legal Amazon has been conducted since 1988. The annual deforestation rates in the region are used by the Brazilian government to establish public policy. The annual deforestation rates are estimated from the deforestation increments identified in each satellite image over the Legal Amazon. The first data presentation is made for December of each year, in the form of an estimate. The consolidated data are presented in the first half of the following year. PRODES uses images from Landsat satellites (30 m of spatial resolution with 16-day revisit). Images from the American satellite LANDSAT-5 TM have historically been the most used by the project, but images from the CCD sensor on board the CBERS-2/2B, satellites of the Sino-Brazilian remote sensing program, have been widely used. PRODES also used LISS-3 images from the Indian IRS-1 satellite and images from the UK-DMC2 satellite. Currently, massive use is being made of Landsat-8 OLI, CBERS-04 and IRS-2 images. The minimum area mapped by PRODES is 6.25 hectares regardless of the instrument used. The PRODES estimates have been considered reliable by national and international scientists with an accuracy level close to 95% (Kintisch, 2007).

In a web page of the **INPE**[2] there is a list of articles in scientific journals that have cited the PRODES Amazon. As of November 25, 2020, PRODES is cited in 1278 articles and in 428 journals (Prodes Amazônia, 2020).

[1]http://www.obt.inpe.br/OBT/assuntos/programas/amazonia/prodes
[2]http://www.obt.inpe.br/OBT/assuntos/programas/amazonia/prodes/citacoes-ao-prodes

6

Remote Sensing of Water

6.1 Homework Solution

6.1.1 Subject

Spatio-temporal monitoring of rainfall from CHIRPS data.

6.1.2 Abstract

The Climate Hazards Group InfraRed Precipitation with Station data (CHIRPS) is a dataset of precipitation data in regions of the Earth with more than 35 years of data time series. The geographic extent of data is available between latitudes 50° S to 50° N and all longitudes from 1981 to the present. The CHIRPS data incorporate a climatological database, satellite imagery at 0.05° spatial resolution, and *in situ* data from meteorological stations to create precipitation time series for trend analysis and seasonal drought monitoring. We aimed to obtain CHIRPS rainfall data to perform spatio-temporal rainfall mapping in the Sulsiani watershed, Mali, Africa. Different CHIRPS precipitation data mapping packages and techniques are evaluated for rainfall monitoring.

Keywords: Drought, mapping, rainfall, R software.

6.1.3 Introduction

Since 1999, scientists have been developing techniques for producing precipitation maps in support of rainfall monitoring mainly in areas where surface rainfall measurement data are scarce. Estimating the spatio-temporal variation of rainfall is a key aspect of early drought warning and environmental monitoring. However, estimates derived from satellite data can be uncertain about the quality of results due to the occurrence of complex terrain variation that can affect the intensity of extreme precipitation events. The CHIRPS data were created in collaboration with scientists at the U.S. Geological Survey (USGS) Earth Resources Observation and Science Center (EROS) in order to provide complete, reliable, and up-to-date datasets for use in different analyses with meteorological applications. The output of the data is determined based on the combination of precipitation prediction models with interpolated data from weather stations concerning relief effects. New features from satellite observations such as NASA and NOAA satellite precipitation estimates are added to the data modeling to develop rainfall climate surfaces at 0.05° resolution. More details on the algorithm used to obtain and validate CHIRPS data quality in hydrological forecasting and trend analysis studies in southeastern Ethiopia can be seen in Funk et al. (2015).

6.1.4 Objective

The objective of the homework is to obtain CHIRPS rainfall data to perform spatio-temporal mapping of rainfall in the Sulsiani watershed in Mali, Africa.

R software is used as an example; however, other software can be used to accomplish this task at the student's discretion.

6.1.5 Enabling packages

The R packages `chirps` (Sousa et al., 2022), `terra` (Hijmans et al., 2022), `sf` (Pebesma et al., 2021), `ggplot2` (Wickham et al., 2022), and `stars` (Pebesma et al., 2022) are required to obtain daily CHIRPS rainfall data, and are enabled for analysis.

```
library(chirps)
library(terra)
library(sf)
library(stars)
library(ggplot2)
```

6.1.6 Import polygon

The polygon used in the analysis refers to a Sulsiani watershed, Mali, Africa. The `st_read` function is used to import the polygon feature with attributes.

```
lim <- st_read("C:/manuscript/Rusle/shape/bsulsiani.shp")
```

The `vect` function is used to convert the `sf data.frame` class file into the `SpatVect` class.

```
v <- vect(lim)
```

6.1.7 Definition of analysis period

An analysis period is set between the beginning and end of the year 2019.

```
dates <- c("2019-01-01","2019-12-31")
```

6.1.8 Collect data

The `get_chirps` function is used to get the rainfall data at the location of interest.

```
a1 <- get_chirps(v, dates, server = "CHC", as.raster = TRUE)
#Fetching data as GeoTIFF files from CHC server
#Getting CHIRPS in a .05 deg resolution
```

6.1.9 Statistical analysis of data

Boxplot analysis is used to graphically explore summary statistics of the data obtained using the `boxplot` function.

```
boxplot(a1)
```

Thereby, it is possible to obtain notions about the seasonality of the data over the studied time period (Figure 6.1).

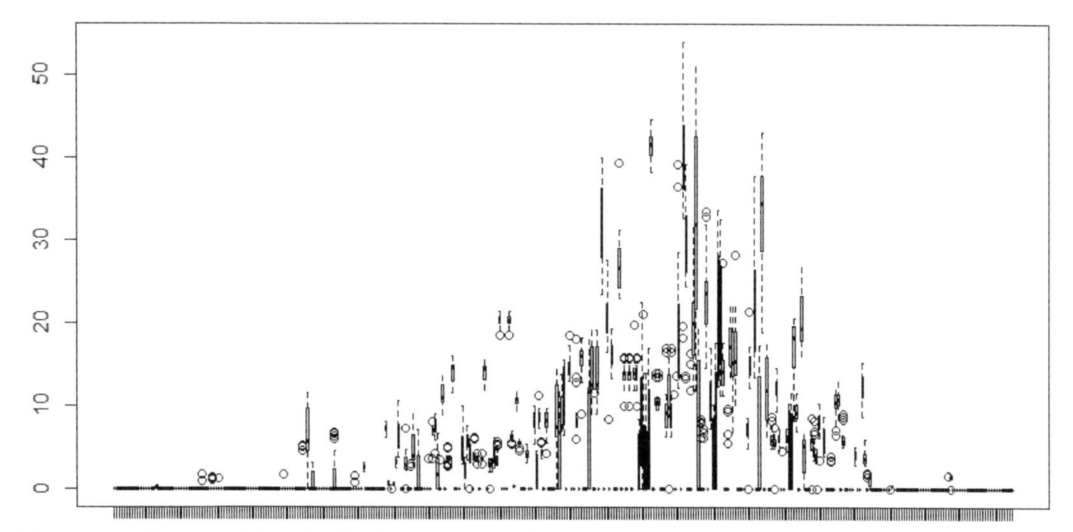

FIGURE 6.1 Boxplot analysis to explore rainfall variation in the Sulsiani watershed, Mali, Africa, in the year 2019.

6.1.10 Export results

The `writeRaster` function is used to export the raster files obtained in the area.

```
terra::writeRaster(a1, "C:/sr/chirps/a1.tif", filetype = "GTiff", overwrite = TRUE)
```

6.1.11 Rainfall mapping

The `read_stars` function is used to import the data into R via the `stars` package.

```
tif <- read_stars("C:/sr/chirps/a1.tif", package = "stars")
```

The results can be mapped in different ways. In this case, we chose to map the results comparatively using the `ggplot` function.

```
ggplot() +
    geom_stars(data = tif) +
     scale_fill_viridis_c(limits = c(0,50)) +
    geom_sf(data = lim, fill = NA) +
   coord_sf(datum=st_crs(4326)) +
    facet_wrap( ~ band) +
    theme_void()
```

With this type of annual data mapping, the visualization of spatial details of rainfall variation is limited; however, it is possible to comparatively verify times of the year with higher rainfall in the watershed (Figure 6.2).

Another simple mapping option can be accomplished with a `plot` from the `stars` package. In this case, the class intervals are defined by the Jenks method.

```
plot(tif, main = NULL, col = grey((5:10)/10), breaks = "jenks")
```

As a geovisualization alternative, the space-time variation patterns of rainfall can be observed in a mapping approach by the `stars` package, in which case the classification of data intervals is established according to the frequency distribution by the Jenks method (Figure 6.3).

After annual analysis of the spatio-temporal mapping, a subset is performed between days 240 to 260 of the data using the bracket operator.

```
tif1 <- tif[,,,240:260]
```

The subset is mapped with the `ggplot` function.

```
ggplot() +
    geom_stars(data = tif1) +
     scale_fill_viridis_c(limits = c(0,50)) +
    geom_sf(data = lim, fill = NA) +
   coord_sf(datum=st_crs(4326)) +
    facet_wrap( ~ band) +
    theme_void()
```

Thereby it is possible to check more details of the spatio-temporal variation of rainfall in the watershed from August 28 to September 17, 2019 (Figure 6.4).

FIGURE 6.2 Mapping the spatio-temporal variation of rainfall in the Sulsiani watershed, Mali, Africa, in the year 2019.

6.2 Solved Exercises

6.2.1 Cite five applications of remote sensing studies of water.

A: Determination of water surface area of streams, rivers, lakes, reservoirs and seas with passive multispectral remote sensing; determination of organic and inorganic constituents of water with multispectral spectroradiometer; determination of water depth by bathymetry with LIDAR and SONAR; determination of water surface temperature with thermal infrared remote sensing; and determination of rainfall with passive RADAR.

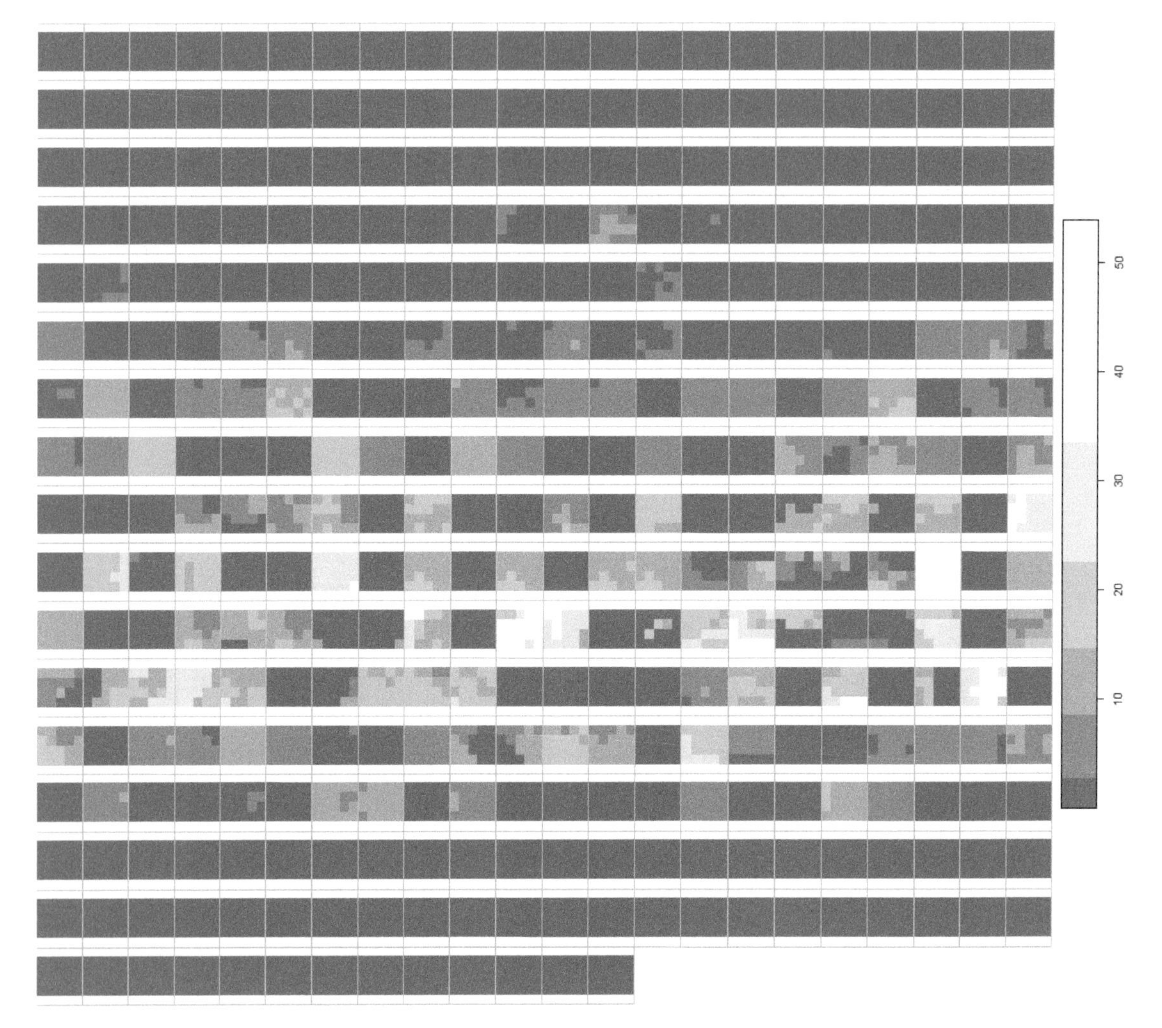

FIGURE 6.3 Mapping spatio-temporal variation of rainfall Sulsiani watershed, Mali, Africa, in the year 2019 with class intervals defined by the Jenks method.

6.2.2 Cite the possible paths of the radiance that reaches a body of water.

A: The total radiance (L_T) recorded from the water by an aircraft- or satellite-borne is a function of the unwanted path radiance (L_P), the reflected radiance from the surface layer of water (L_S), the reflected radiance from the volumetric subsurface layer of water (L_V), and the reflected radiance from the bottom of the body of water (L_B).

6.2.3 Cite four components of the volumetric radiance of water.

A: The subsurface volumetric radiance leaving the water column in the direction of the sensor is a function of the concentration of clear water (w), inorganic suspended sediments (SM), organic chlorophyll a (Chl), dissolved organic material (DOM), and the total amount of attenuation by scattering in the water column in each of these constituents ($c\lambda$).

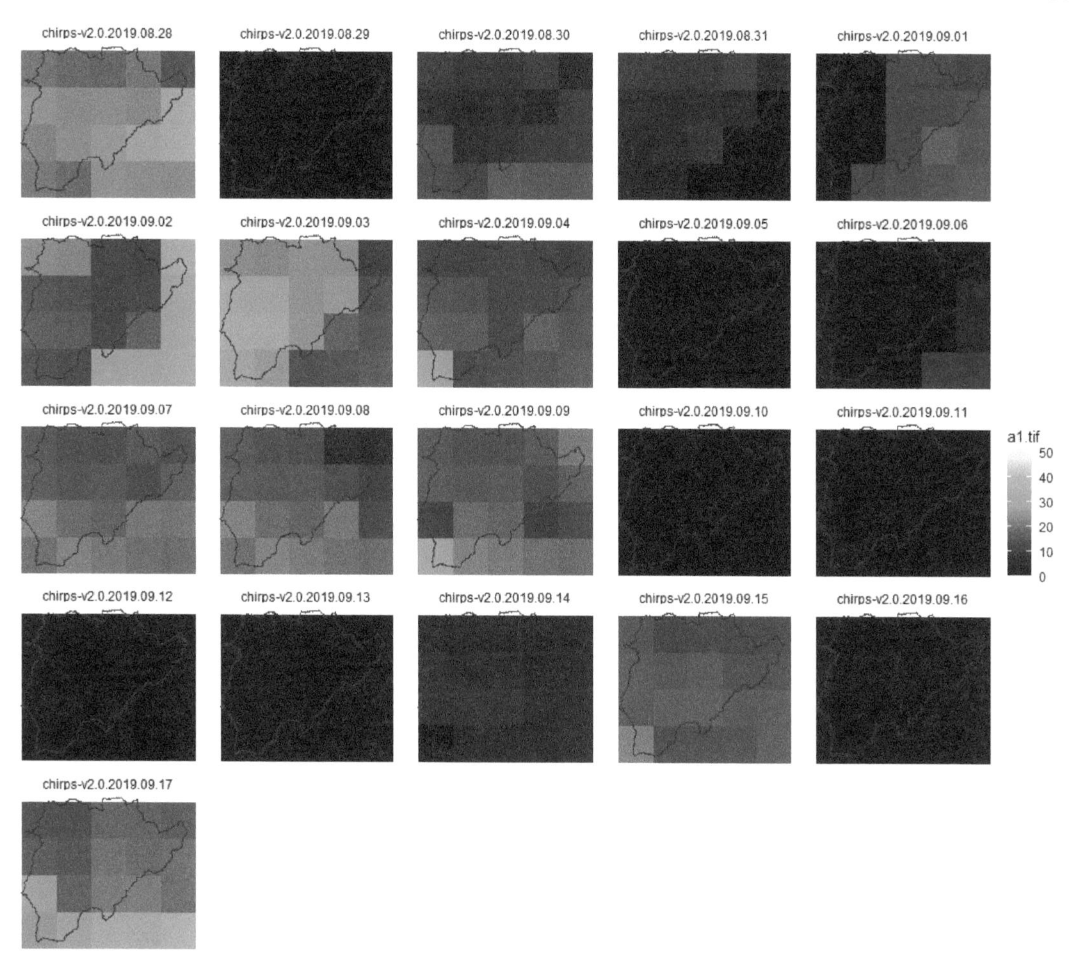

FIGURE 6.4 Mapping the spatio-temporal variation in rainfall Sulsiani watershed, Mali, Africa, from August 28 to September 17, 2019.

6.2.4 Explain how water vapor monitoring can be done in South America and how this information can be used in practice.

A: Atmospheric high-level water vapor can be studied in South America by GOES-East imagery at wavelength 6.19 μm, channel 8, at spatial resolution of 2 km.

6.2.5 Explain how a spectral band ratio remote sensing moisture index can be used to separate water from terrain and to detect the moisture content of the monitored object.

A: The normalized difference water index (NDWI), based on the ratio of (green - NIR)/(green + NIR) bands, can be used to separate water from terrain so that high NDWI values are used to map the water body and separate it from the terrain. Normalized difference water index 2 (NDWI2), on the other hand, based on the ratio of bands (NIR - SWIR2)/(NIR + SWIR2), can be used to detect the moisture of objects in the terrain, so the higher the NDWI2, the higher the moisture of the object analyzed in the image.

7

Remote Sensing of Soils, Rocks, and Geomorphology

7.1 Homework Solution

7.1.1 Subject

Monitoring spatial variation of relief for determining morphometric terrain variables from SRTM data.

7.1.2 Abstract

The Shuttle Radar Topography Mission (SRTM) is an international research effort that obtained digital elevation models on a near-global scale from 56° S to 60° N to generate the most comprehensive high-resolution digital topographic database of the Earth in 2000, prior to the release of ASTER GDEM in 2009. However, SRTM uses RADAR observations to construct the digital elevation model (DEM) while ASTER uses stereo imagery and photogrammetric techniques to extract the DEM. The DEM obtained in the computational practice is used to determine morphometric terrain variables from mathematical modeling methodology considering topological features of relief variation in a locality in state of Bahia, Brazil.

Keywords: Mapping, morphology, relief, software R.

7.1.3 Introduction

A digital elevation model is a representation of the topographic surface of the Earth (bare earth) excluding trees, buildings, and any other surface objects. A DEM can be created from a variety of sources, such as from a topographic map. The Shuttle Radar Topography Mission digital elevation data comprise an international research effort that has obtained digital elevation models on a near-global scale. With the SRTM V3 (SRTM Plus) product provided by NASA JPL, it is possible to use a DEM with a spatial resolution of 1 arc-second ($\sim$ 30 m). The method of collecting SRTM data is known as Synthetic Aperture Radar (SAR) interferometry. By this method, two SAR antennas collected radar data separated by a 60-meter extension arm and interferometric software was used to generate topographic data. SRTM data can help scientists better understand the Earth's systems of land, water, air, and life, how they interact, and how they are changing (Farr et al., 2007).

7.1.4 Objective

The objective of the homework is to obtain SRTM data to perform spatial mapping and determination of slope, aspect and hillshade variables in a rural property in the municipality of Correntina, state of Bahia, Brazil.

The software R is used as an example; however, other software can be used to accomplish this task at the student's discretion.

7.1.5 Enabling packages

The `library` function is used in the R console to enable the R packages `raster` (Hijmans et al., 2020), `RStoolbox` (Leutner et al., 2019), `rgdal` (Bivand et al., 2021), and `terra` (Hijmans et al., 2022).

```
library(raster)
library(RStoolbox)
library(rgdal)
library(terra)
```

7.1.6 Obtain SRTM data on the Internet

The digital elevation model of the area of interest, that is used in the computational practice, can be obtained for **download**[1].

7.1.7 Importar dados no R

The terrain slope, aspect and shading variables are determined from active RADAR remote sensing data obtained from the SRTM mission. The digital elevation model was obtained from the `AppEEARS` platform at x, y spatial resolution of 0.0002777778°, 0.0002777778°, respectively. The remote sensing monitoring data is imported into R with the `rast` function.

```
dem<-rast("C:/sr/c6/SRTMGL1_NC.003_SRTMGL1_DEM_doy2000042_aid0001.tif")
```

7.1.8 Transform data projection

The `project` function is used to convert the digital elevation model to the Universal Transverse Mercator (UTM) map projection, zone 23 South, with x, y spatial resolution of 29.3184 by 29.3184 m, respectively.

```
utm23s<-"+proj=utm +zone=23 +south +datum=WGS84 +units=m +no_defs"
demprj <- terra::project(dem, crs(utm23s))
```

[1] http://www.sergeo.deg.ufla.br/sr/downloads/srtmprj.zip

7.1.9 Convert to raster

The image with altitude data in the SpatRaster class is converted to the raster class with the raster function.

```
demr <- raster(demprj)
```

7.1.10 Determining terrain morphometric variables

The variables slope, aspect, and hillshade are determined with the terrain and hillshade functions.

```
demSlope <- terrain(demr, opt="slope")
demAspect <- terrain(demr, opt="aspect")
demHill <- hillShade(demSlope, demAspect, 40,270)
```

7.1.11 Perform mapping

The results are stacked with the stack function, renamed with the names function, and mapped with the plot function.

```
relief<-stack(demr, demSlope, demAspect, demHill)
names(relief)<-c("altitude", "slope", "aspect", "hillshade")
plot(relief, col=grey(0:100/100))
```

With the slope and hillshade maps it is possible to verify that the central region of the area where there was flooding of the area with the formation of the dam presented less rugged relief with a funnel shape (Figure 7.1)

7.2 Solved Exercises

7.2.1 List five applications of remote soil sensing and explain a situation where monitoring may be limited.

A: Identify drainage network, geomorphology, watershed, slope, and aspect with active remote sensing by synthetic interferometry RADAR. Monitoring can be limited when necessary to remove sinks by image processing from the digital elevation model in order to improve morphometric and hydrological characterizations.

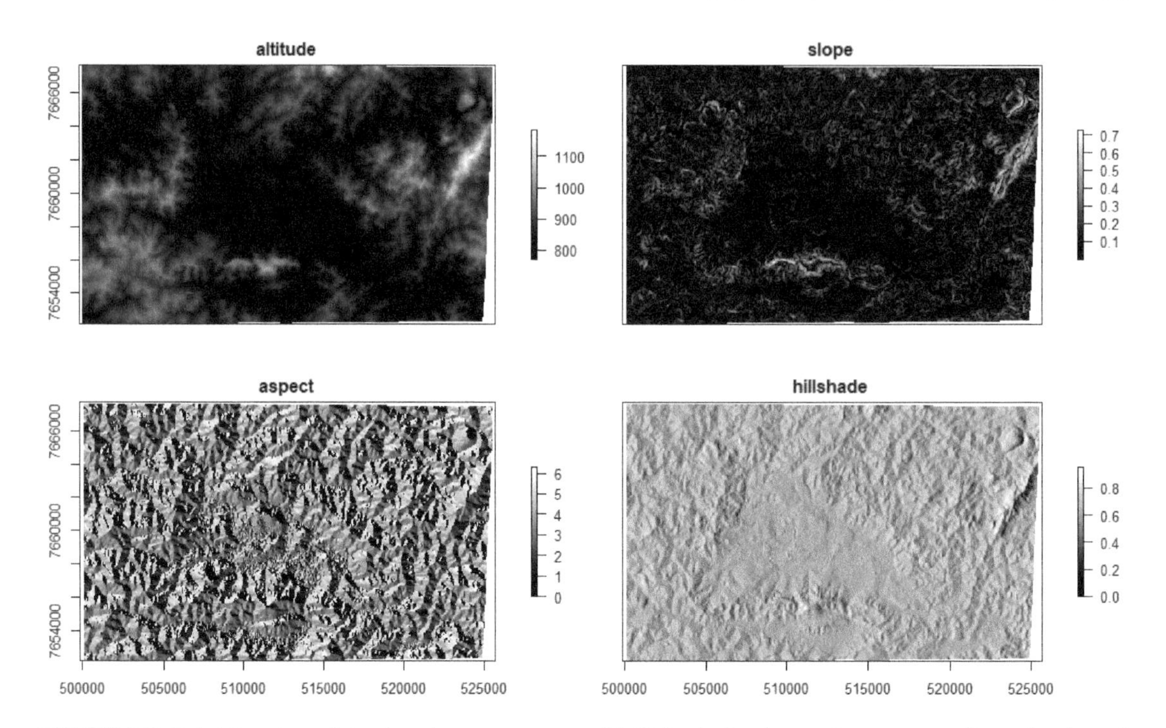

FIGURE 7.1 Mapping altitude, slope, aspect and hillshade in the region surrounding the funnel dam.

7.2.2 Explain the possible pathways of radiance reaching exposed soil.

R: The total radiance coming from an exposed soil and recorded on a sensor on board a satellite (LT) is the sum of L_P, the downward radiance from the Sun and atmosphere that never reaches the ground surface but is recorded by the sensor, L_S, the part of the direct and diffuse solar radiation that reaches the air-soil interface and penetrates to approximately half wavelength depth into the soil and emerges from the soil column by reflection and by scattering for recording by the sensor, and L_V, the part of the direct and diffuse solar radiation that penetrates a few millimeters to centimeters into the soil column and is recorded by the sensor.

7.2.3 List six characteristics that can interfere with the variation in the reflectance of exposed soil.

A: Soil spectral reflectance characteristics vary with texture, moisture content, organic matter content, iron oxide content, soil salinity, and ground surface roughness.

7.2.4 Explain how soil monitoring can be done based on a digital elevation model and two variables that can be obtained from altitude.

A: A LIDAR survey can be done in the area of interest to determine the digital elevation model of the area and then calculate slope and aspect from altitude.

7.2.5 Explain how to calculate a remote sensing index to study bare soil and what the practical interpretation of this index is.

A: The Bare Soil Index (BSI) can be calculated by the equation:

$$BSI = \frac{(SWIR3 + Red) - (NIR + Blue)}{(SWIR3 + Red) + (NIR + Blue)} \tag{7.1}$$

where Blue, NIR and SWIR3 are the blue, near-infrared and shortwave infrared bands, respectively. The higher the BSI, the greater the possibility of bare soil occurring in the studied region.

7.2.6 Which index is not used for soil monitoring with passive remote sensing?

a. SAVI.
b. BSI.
c. SRTM. [X]
d. BSCI.
e. None of the alternatives.

7.2.7 Which mission used to monitor relief for soil studies with active remote sensing?

a. SAVI.
b. BSI.
c. SRTM. [X]
d. BSCI.
e. None of the alternatives.

7.2.8 Raster data for an area of interest is obtained from the AppEEARS platform with 82424 pixels of altitude at 30-m spatial resolution. Based on this information, what is the size of the total raster area in hectares?

A: The total raster area is 7418.16 ha.

```
resol<-30*30
area<-82424*resol
areaha<-area/10000
areaha
#[1] 7418.16
```

7.2.9 What are the water absorption wavelengths in the reflectance curve of moist sandy soil?

A: The higher the moisture content in sandy and clay soil, the lower the reflectance in the visible and near-infrared, especially in the water absorption bands at 1.4, 1.9, and 2.7 μm.

7.2.10 Determine the textural class of the soil in triangle classification.

Soil texture is classified in this exercise according to the Brazilian classification system (Santos et al., 2018). The data are obtained based on soil sampling result data, in the 0 to 20 cm layer, of clay, silt, sand (%) and organic matter quintiles (g kg^{-1}) in a mesh of 67 georeferenced vertices in coffee plantation in Cafua farm in Ijaci, state of Minas Gerais, Brazil (Figure 7.2).

```
# Enable the package
library(soiltexture)
# Draw the triangle of the Brazilian textural classification
TT.plot(class.sys = "SiBCS13.TT", class.p.bg.col = c("red", "green", "blue", "pink",
                                                    "purple"))
# Create a table of soil texture data for a coffee farm where clay, silt and sand
# are in % and organic matter (OC) in gkg-1
cafua <- data.frame(
"CLAY" = c(66,62,74,57,55,56,51,52,57,51,56,46,53,50,48,50,55,59,57,52,46,48,48,46,
          53,64,75,72,57,46,50,51,56,50,47,48,49,42,46,58,66,70,70,73,53,48,42,47,
          59,66,69,70,73,58,59,68,67,73,58,54,47,47,66,58,72,67,69),
"SILT" = c(11,11,14,10,9,10,9,11,8,10,8,11,10,8,12,9,10,6,4,6,8,5,6,5,14,14,13,13,
          10,10,9,8,8,10,10,9,7,11,8,13,14,15,15,16,9,11,18,9,13,14,16,16,16,16,
```

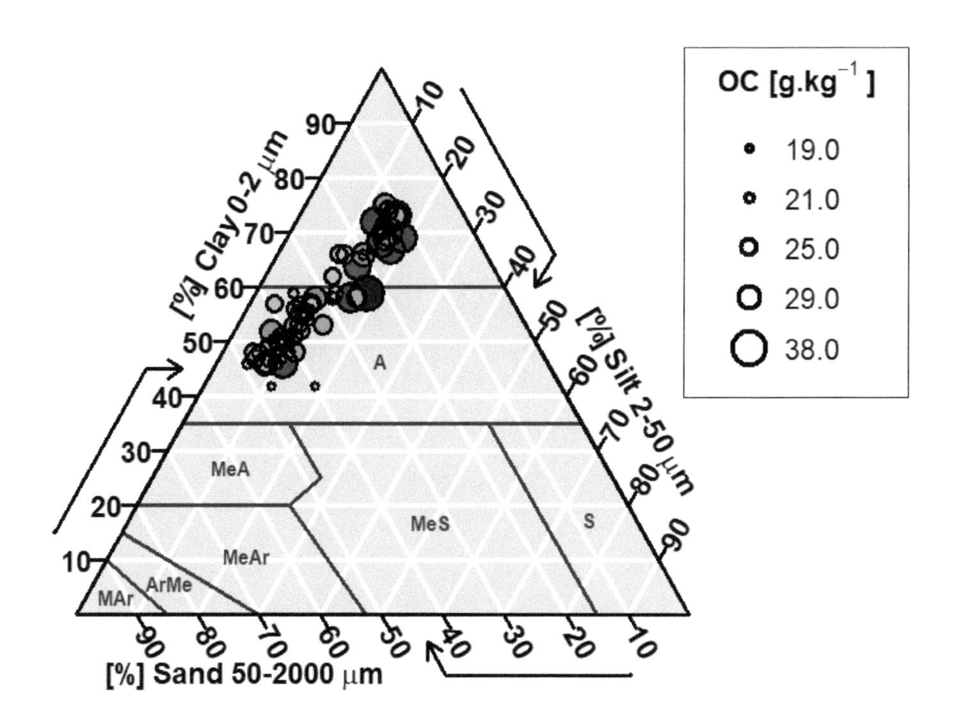

FIGURE 7.2 Determination of textural class and organic carbon in the 0-20 cm layer of soil obtained from a sampling grid in a coffee plantation.

```
            18,16,18,16,10,11,6,8,10,17,14,18,19),
"SAND" = c(23,27,12,33,36,34,40,37,35,39,36,43,37,42,40,41,35,35,39,42,46,47,46,49,
            33,22,12,15,33,44,41,41,36,40,43,43,44,47,46,29,20,15,15,11,38,41,40,44,
            28,20,15,14,11,26,23,16,15,11,32,35,47,45,24,25,14,15,12),
"OC" = c(26,25,27,24,26,21,21,25,24,26,25,34,25,25,27,25,25,21,25,29,29,26,24,21,27,
          33,29,34,27,21,21,26,19,21,21,21,25,19,26,22,21,21,25,27,26,21,19,19,24,29,
          29,34,29,34,38,34,34,34,29,21,26,25,25,27,21,34,34))
# Show the table
cafua
# Determine soil texture in textural triangle and organic matter in symbology
# Circular quartiles
TT.plot(class.sys = "SiBCS13.TT", tri.data = cafua, z.name = "OC",
main = "Soil Texture")
# Calculate summary statistics and quartile of organic matter
z.cex.range <- TT.get("z.cex.range")
def.pch <- par("pch")
def.col <- par("col")
def.cex <- TT.get("cex")
oc.str <- TT.str(cafua[, "OC"], z.cex.range[1], z.cex.range[2])
# Legend
legend(x = 99, y = 90, title = expression( bold('OC [g.kg'^-1 ~ ']') ),
legend = formatC(c(min( cafua[, "OC"] ),
quantile(cafua[,"OC"] ,probs=c(25,50,75)/100),
max( cafua[, "OC"] )),
format = "f", digits = 1, width = 4, flag = "0"),
pt.lwd = 4, col = def.col, pt.cex = c(min(oc.str),
quantile(oc.str ,probs=c(25,50,75)/100), max(oc.str)),
pch = def.pch, bty = "o", bg = NA, text.col = "black", cex = def.cex)
```

7.2.11 Map soil organic carbon at different depths and different color options in Brazil.

7.2.11.1 Obtain soil organic carbon maps of Brazil on the Internet.

- **EMBRAPA**[2]

Organic carbon data at depths 0-5, 5-15, and 15-30 cm, are imported and stacked for mapping (Figure 7.3).

```
# Enable the package
library(raster)
# Import raster with soil organic carbon at different depths
co05 <-raster("C:/sr/c6/Brazil/CO_0_5cm_BR.tif")
co15 <-raster("C:/sr/c6/Brazil/CO_5_15cm_BR.tif")
co30 <-raster("C:/sr/c6/Brazil/CO_15_30cm_BR.tif")
# Perform soil organic carbon data stacking
co<-stack(co05, co15, co30)
```

[2]https://www.embrapa.br/busca-de-noticias/-/noticia/2062813/solo-brasileiro-agora-tem-mapeamento-digital

```
# Map the soil organic carbon
plot(co, col = rev(rainbow(100)))
```

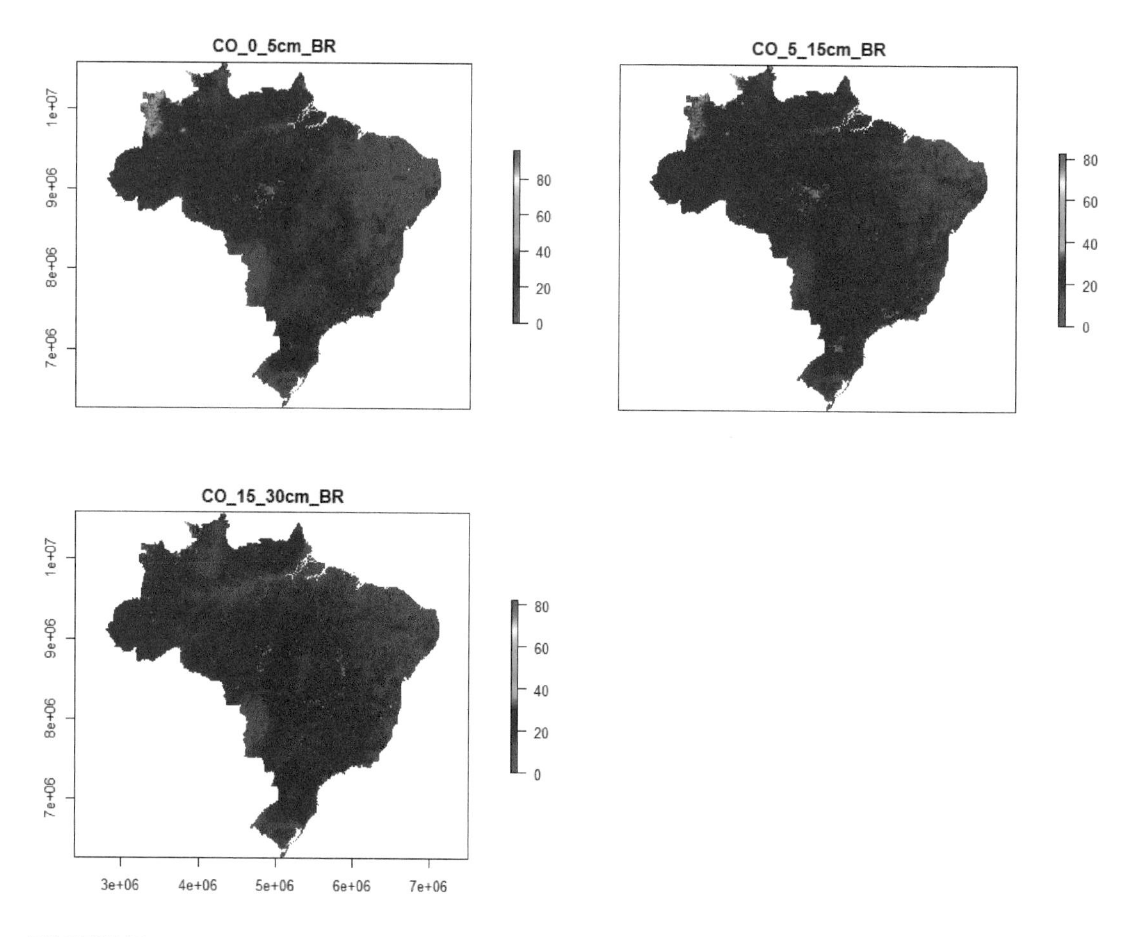

FIGURE 7.3 Mapping organic carbon at different depths in Brazil.

7.2.12 Map the soil classes of Brazil in order level.

7.2.12.1 Obtain the soil class chart of Brazil on the Internet.

The obtained information plan represents the geographical distribution of soils in Brazil, according to the Brazilian Soil Classification System (SiBCS, 2006), classified up to the third categorical level, at scale 1:5,000,000.

- **GEOINFO**[3]

The soils of Brazil are mapped in order level obtained from the EMBRAPA Solos database (Figure 7.4).

[3]http://geoinfo.cnps.embrapa.br/layers/geonode%3Abrasil_solos_5m_20201104

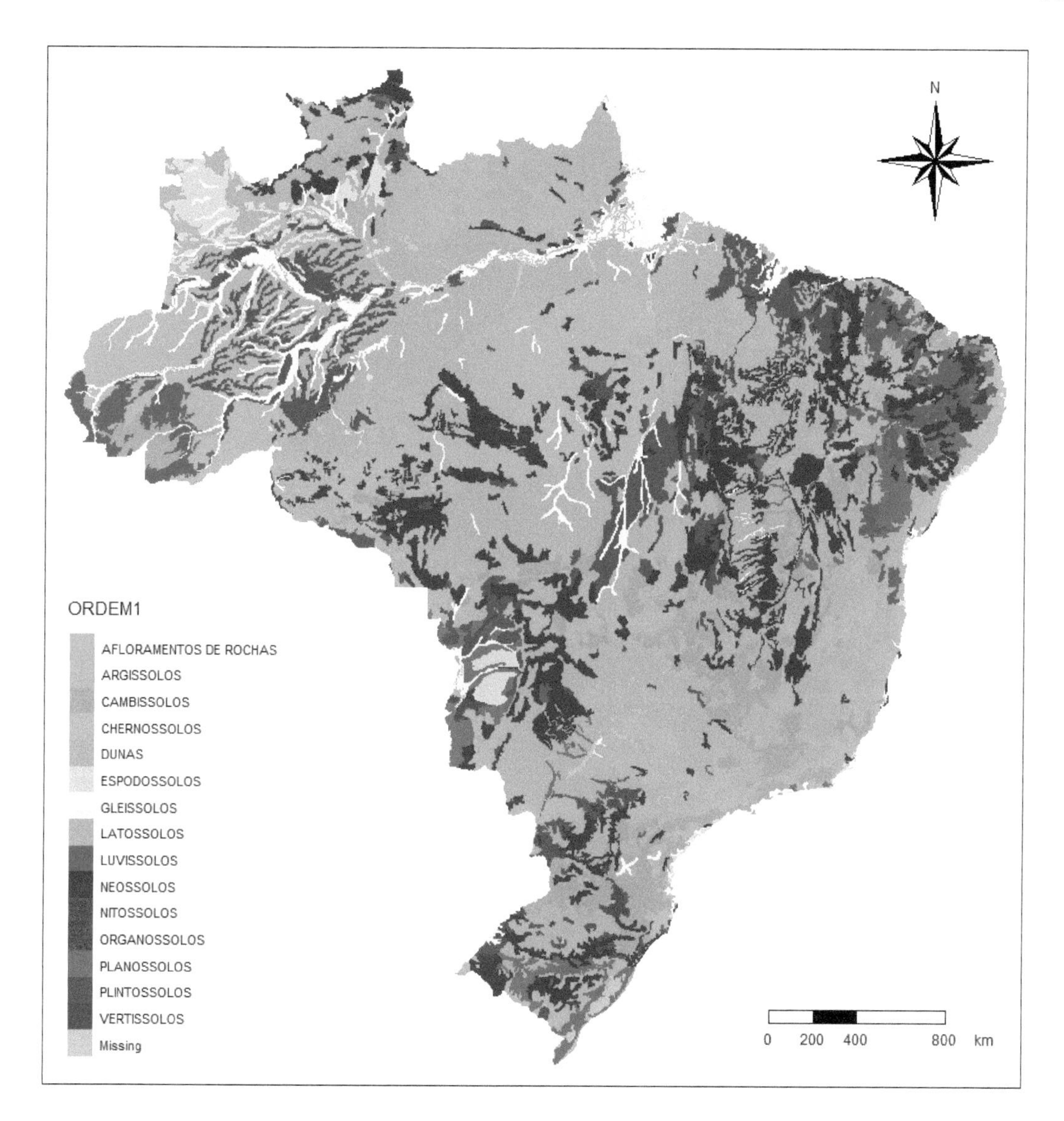

FIGURE 7.4 Soil mapping in Brazil at the order level.

```
# Enable the packages
library(raster)
library(sf)
library(tmap)
# Import feature with soil mapping
soils <- st_read("C:/sr/c6/brasil_soils_5m_20201104/brasil_soils_5m_20201104.shp")
names(soils) # Get the name of the table columns
# Map
order= tm_shape(soils[7:7]) +
   tm_fill("ORDER1",palette="Accent") +
  tm_borders(alpha=0) +
  tm_legend(position = c("left", "bottom"),
    frame = FALSE,
```

```
    bg.color="white") +
  tm_compass(type = "8star", position = c("right", "top")) +
  tm_scale_bar(breaks = c(0, 200, 400, 800), text.size = 1,
               position = c("right", "bottom")) +
  tmap_options(check.and.fix = TRUE)
order
```

8

Remote Sensing of the Atmosphere

8.1 Homework Solution

8.1.1 Subject

Planetary-scale cloud fraction monitoring with Sentinel-5P remote sensing data.

8.1.2 Abstract

Cloud fraction monitoring can be performed on a planetary scale with Sentinel-5P remote sensing data. Obtaining cloud properties with the Sentinel-5P satellite's TROPOMI sensor is based on the OCRA and ROCINN algorithms currently in use in the GOME and GOME-2 operational products. Sentinel-5P monitoring data is used as an example to establish a processing and mapping flow of administrative boundaries and atmospheric gases at a location of interest. Thereby it is possible to visualize higher cloud fraction between June 1 and 2, 2019 in shades tending towards red on a planetary scale forming patterns with global wind belts.

Keywords: Cloud, mapping, `rgee`, satellite, TROPOMI.

8.1.3 Introduction

Thick, continuous cloud cover forms a significant barrier to radiation penetration. The radiation reflected by clouds depends on the amount, cover, and thickness of clouds. The proportion of incident radiation that is reflected is called the "albedo". The type of cloud affects the albedo. Observations made at the surface tend to be 10% higher than satellite estimates due to the observer's perspective (Barry & Chorley, 2010).

Obtaining cloud properties with the TROPOMI sensor on the Sentinel-5P satellite is based on the OCRA and ROCINN algorithms currently in use in the GOME and GOME-2 operational products. With OCRA it is possible to obtain the cloud fraction using spectral region measurements in the ultraviolet and visible regions, and ROCINN obtains the cloud height (pressure) and optical thickness (albedo) using measurements in the oxygen A-band at 760 nm. Version 3.0 of the algorithms is used in order to exploit more realistic treatment of clouds as optically uniform layers of light-scattering particles. In addition, cloud parameters are also provided for a cloud model that assumes the cloud as a Lambertian reflector. Further information on cloud characteristics that can be obtained using the TROPOMI sensor can be accessed at the website **TROPOMI**[1].

[1] http://www.tropomi.eu/data-products/cloud

8.1.4 Objective

The objective of the homework is to monitor the spatial variation of clouds on a planetary scale using remote sensing data obtained by the TROPOMI sensor onboard the Sentinel-5P satellite.

8.1.5 Data acquisition

Sentinel-5P monitoring data can be accessed in the cloud via R and Earth Engine (EE). The R packages `geobr` (Pereira et al., 2021), `sf` (Pebesma et al., 2021), `rgee` (Aybar et al., 2022), and `rmapshaper` (Teucher et al., 2022) are used as an example to establish a stream for processing and mapping administrative boundaries and atmospheric gases at a location of interest, in this case Brazil. The packages are enabled with the `library` function.

```
library(geobr)
library(sf)
library(rgee)
library(rmapshaper)
```

The `ee_Initialize` function is used to initialize EE.

```
ee_Initialize(drive = TRUE)
```

The boundaries of Brazilian states are obtained using the `read_contry` function of the `geobr` package. The Brazilian boundary data is obtained at scale 1:250,000, using the geodetic reference system SIRGAS2000, EPSG 4674, and 1.4 MB in size.

```
br<-read_country(year = 2020, simplified = TRUE, showProgress = TRUE)
#Using year 2020
#Downloading: 1.4 MB
```

The `st_transform` function of the `sf` package is used to transform the projection of the data from SIRGAS2000 to WGS-84, using the reference coordinate system with EPSG 4326.

```
brwgs<-st_transform(br, crs = 4326)
```

The `object.size` function is used to evaluate the file size, in this case 2073640 bytes.

```
object.size(brwgs)
#2073640 bytes
```

The `ms_simplify` function from the `rmapshaper` package is used to reduce the file size since this type of mapping does not require much vertex detail. The file is transformed into the `sp` class and then transformed back into the `sf` class using the pipe operator.

```
simplepolys <- rmapshaper::ms_simplify(input = as(brwgs, 'Spatial')) %>%
  st_as_sf()
```

This resulted in an object size of 149808 bytes. This object with the Brazilian boundaries is used as a reference in mapping the atmospheric variation of cloud fraction.

```
object.size(simplepolys)
#149808 bytes
```

The `sf_as_ee` function is used to convert vectors to multipolygons of class `ee`.

```
polsee <- simplepolys  %>% sf_as_ee()
```

The polygon fill is removed by using a chunk of `ee$Image()$byte()` code associated with the `paint` argument. This makes it easier to visualize Sentinel-5P imagery data inside the polygons mapped over the image.

```
empty <- ee$Image()$byte()
outline <- empty$paint(featureCollection=polsee,color=1,width=3)
```

A search period for images in the Earth Engine data collection is set between June 1 and June 6, 2022.

```
startDate = '2019-06-01'
endDate = '2019-06-02'
```

The `ee$ImageCollection` function is used to obtain Sentinel-5P imagery data for the proposed period. The filtered file consisted of 14 images in the size of 390.96 GB and a set of variables about cloud fraction, cloud top pressure, cloud top height, cloud base pressure, cloud base height, cloud optical depth, surface albedo, sensor azimuth angle, sensor zenith angle, solar azimuth angle, solar zenith angle.

```
collection <- ee$ImageCollection('COPERNICUS/S5P/OFFL/L3_CLOUD')$
  filterDate(startDate, endDate)
```

The filtered results are displayed with the `ee_print` function.

```
ee_print(collection)
#--------------- Earth Engine ImageCollection --
#ImageCollection Metadata:
# - Class                          : ee$ImageCollection
# - Number of Images               : 14
# - Number of Properties           : 26
# - Number of Pixels*              : 9.9792e+10
# - Approximate size*              : 390.96 GB
#Image Metadata (img_index = 0):
# - ID       : COPERNICUS/S5P/OFFL/L3_CLOUD/20190601T011830_20190607T004228
# - system:time_start              : 2019-06-01 01:40:04
# - system:time_end                : 2019-06-01 02:38:28
# - Number of Bands                : 11
# - Bands names                    : cloud_fraction cloud_top_pressure cloud_top_height
#cloud_base_pressure cloud_base_height cloud_optical_depth surface_albedo
```

```
#sensor_azimuth_angle sensor_zenith_angle solar_azimuth_angle solar_zenith_angle
# - Number of Properties        : 28
# - Number of Pixels*           : 7.128e+09
# - Approximate size*           : 27.93 GB
#Band Metadata (img_band = 'cloud_fraction'):
# - EPSG (SRID)                 : WGS 84 (EPSG:4326)
# - proj4string                 : +proj=longlat +datum=WGS84 +no_defs
# - Geotransform                : 0.01 0 -180 0 0.01 -70
# - Nominal scale (meters)      : 1113.195
# - Dimensions                  : 36000 18000
# - Number of Pixels            : 6.48e+08
# - Data type                   : DOUBLE
# - Approximate size            : 2.54 GB
# ------------------------------------------------------------------------------
# NOTE: (*) Properties calculated considering a constant  geotransform and data type.
```

8.1.6 Mapping

Parameters for visualization and selection of bands of interest are defined for mapping results centered in Mato Grosso, Brazil. A color palette of 'black', 'blue', 'purple', 'cyan', 'green', 'yellow', 'red' in the hex system is used for visualization of cloud fraction density in the atmospheric column. The `Map$addLayer` function is used to map the average of monitored cloud fraction values in the period with the `$mean()` function.

```
palette <- c('#000000', '#0000FF', '#800080', '#00FFFF', '#00FF00', '#FFFF00', '#FF0000')
imageVisParam <- list(bands = c('cloud_fraction'), min = 0, max = 0.95,
                      palette= palette)
Map$setCenter(-51.85547,-12.12526,4)
Map$addLayer(collection$mean(), imageVisParam, 'S5P Cloud', legend = T)+
Map$addLayer(outline, list(palette='white'),'Brazil')
```

Afterward, there is the possibility to visualize a greater fraction of clouds in shades tending to red on a planetary scale forming patterns with global wind belts. Some locations in the Pacific and Atlantic oceans, as well as in the interior of the Brazilian territory, in parts of the states of São Paulo, Mato Grosso do Sul, Mato Grosso, and Amazonas, presented clear situations with no clouds (Figure 8.1).

8.2 Solved Exercises

8.2.1 How can remote sensing of atmosphere be useful to society?

A: Atmospheric variation can determine radiation imbalances that occur from natural processes, such as astronomical effects of incident shortwave solar radiation, volcanic eruption, aerosols, and human action that introduces greenhouse gases into the atmosphere. From knowledge about

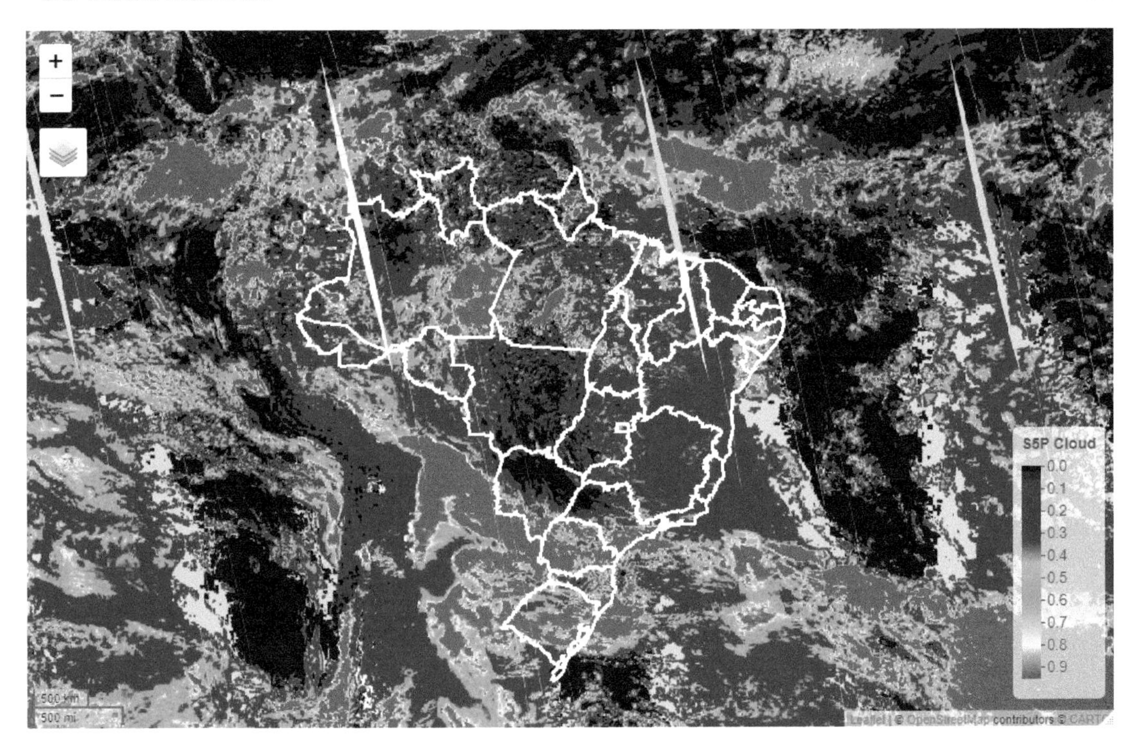

FIGURE 8.1 Mean spatial variation of cloud fraction from Sentinel-5P data around Brazil between June 1 and 2, 2019.

atmospheric functioning, we can contribute to the understanding of climatic phenomena and global climates.

8.2.2 What is the mean composition of the dry atmosphere below 25 km?

a. 99% nitrogen and oxygen. [X]
b. 99% hydrogen and oxygen.
c. 99% hydrogen and ozone.
d. 99% carbon dioxide and ozone.
e. None of the alternatives.

8.2.3 Which passive remote sensing mission used to monitor the concentration of trace gases and aerosols that affect air quality and climate is obtained by satellite?

a. Landsat-8 TIRS.
b. TRMM.
c. Sentinel-5P. [X]
d. SRTM.
e. None of the alternatives.

8.2.4 Which R package specializes in processing and mapping Sentinel-5P data in a region of interest?

a. `gstat`.
b. `S5Processor`. [X]
c. `glcm`.
d. `geoR`.
e. None of the alternatives.

9

Remote Sensing and Digital Image Processing for Project Design

9.1 Homework Solution

9.1.1 Subject

Elaboration of a project abstract in the area of remote sensing.

9.1.2 Abstract

The project title, abstract and keywords are elaborated as follows:

Insights for determining water depth and surface area of hydroelectric power plant dam using remote sensing big data

Marcelo de Carvalho Alves - Federal University of Lavras, Department of Agricultural Engineering, email: *marcelo.alves@ufla.br*

Luciana Sanches - Federal University of Mato Grosso, Department of Sanitary and Environmental Engineering, email: *lsanches@hotmail.com*

Abstract

Brazil has a large water supply used for various purposes, such as transportation, tourism, sports, aquaculture, and power generation. Big data from passive and active remote sensing from multispectral monitoring at different times can be used to characterize the water surface area and conduct bathymetric surveys of the Funil reservoir in Minas Gerais, Brazil. The objective of this project is to use Sentinel-1, Landsat-8, SRTM, and ASTERM GDEM data to determine the water surface area and depth of the Funil dam. Landsat-8 and Sentinel-1 multispectral imaging, product GRD, from June 28, 2021 will be used to comparatively define the surface water area of the dam. The defined water surface area will be used to mask altitude data from the digital elevation models, SRTM and ASTER GDEM, for the period before and after the dam was built. With the use of remote sensing big data it is hoped to provide insights into the determination of depth and water surface area at the Funil dam and to evaluate the use of RADAR image processing and spatial data science techniques to obtain a monitoring routine for the dam based on recent remote sensing data.

Keywords: ASTER GDEM, digital elevation model, Landsat-8, multispectral, RADAR, Sentinel-1, SRTM.

9.2 Solved Exercises

9.2.1 Cite the topics used in the preparation of a scientific project in the area of remote sensing.

A: Title; Authors; Introduction; Hypothesis; Objectives, Methodology; Expected Results; Technology Dissemination; Team; Budget; References.

9.2.2 Cite applications of remote sensing that can be evaluated in scientific research.

A: Remote sensing is used in many applications, such as: medical image analysis, nondestructive evaluation of products on an assembly line, and land resources analysis. In land resources analysis, sensing is useful in modeling the global carbon cycle, ecosystem biology and biochemistry, aspects of global water and energy cycles, climate variability and weather prediction, atmospheric chemistry, solid ground characteristics, population estimates, land use change monitoring, and natural disasters.

9.2.3 In preparing the scientific project in remote sensing, in which section will the project's data analysis be described?

a. Methodology. [X]
b. Conclusions.
c. Bibliographical references.
d. Objectives.
e. Hypotheses.

10

Visual Interpretation and Enhancement of Remote Sensing Images

10.1 Homework Solution

10.1.1 Subject

Determination of remote sensing variables and mapping of RGB color compositions from MODIS imagery.

10.1.2 Abstract

MODerate Imaging Spectroradiometer sensor data are used in various environmental applications. Using MODIS Surface Reflectance products, it is possible to obtain estimates of surface spectral reflectance to determine remote sensing big data in watersheds. We aimed to obtain MODIS imagery data and define a workflow for determining remote sensing variables and mapping the data in the Upper and Middle Paraguay watershed, state of Mato Grosso, Brazil. A MODIS Land Surface Reflectance - V.6.0 (MOD09GA) imagery product referring to Julian day 224, tile h12v10, from August 13, 2018, is obtained for analysis. With the use of RGB color compositions, determination of spectral indices and tasseled cap transformation, it was possible to recognize the real nature of the targets in the watershed, and possible to detect locations with bare soil, water, water associated with riparian forest, urban areas, and tilled and pasture regions in the watershed.

Keywords: Big data, image, mapping, MODerate Imaging Spectroradiometer processing, satellite.

10.1.3 Introduction

The MODerate Imaging Spectroradiometer (MODIS) is used in regional and global studies of the environment, including applications for studies in agriculture, hydrosphere, atmosphere, socio-economics, deforestation, and wildfires. MODIS was conceived by a team of interdisciplinary scientists with extensive experience in operating and using orbital sensor data for data collection, calibration and processing, as well as specifying data and products used to understand global change processes from satellite monitoring (Rudorff et al., 2007).

In Brazil, several applications of MODIS imagery have been developed, including simulations based on Advanced Very High Resolution Radiometer (AVHRR) and Thematic Mapper (TM) sensors. The Brazilian experience with MODIS data ranges from reception, storage, image processing and generation of products in several environmental applications. Free available products are related to terrestrial ecosystems, ocean, and atmosphere (Huete et al., 1999; Lobser & Cohen, 2007; Rudorff et al., 2007).

Using MODIS Surface Reflectance products it is possible to obtain surface spectral reflectance estimation as if it were measured at ground level, in the absence of atmospheric scattering or absorption. The originally obtained low-level processing data are corrected for atmospheric gases and aerosols. In the MOD09GA version 6.1 product, it is possible to obtain bands 1 to 7, in the sinusoidal projection, including 500-m reflectance values, 1-km observation and geolocation statistics (USGS, 2021).

10.1.4 Objective

The objective of this homework is to obtain MODIS imagery data and define a workflow for determining remote sensing variables and mapping the data at a site of interest.

10.1.5 Obtaining imagery by web browsing

Although it is possible to go to the Earth Explorer website and browse for images on the site, it is necessary to register in the data catalog to obtain the data files for analysis with R. The Earth Explorer website is accessed at **EarthExplorer**[1].

10.1.6 Log in to the USGS website

The login page of the Earth Explorer website can be accessed at **login**[2].

10.1.7 Choose image from catalog

A practical solution for starting remote sensing applications is to use MODIS sensor data for an evaluation of an object or process in the region chosen for a particular practical work. In this case, to obtain the information on surface reflectance, you should use the option "MODIS Land Surface Reflectance - V.6.0" (MOD09GA) and download the files instantly in `hdf` format. The option "MODIS Land Surface Reflectance - V.6.1" can also be used at the student's discretion, which is the updated version. The MOD09GA Version 6 product provides an estimate of the Earth surface spectral reflectance of the Moderate Resolution Imaging Spectroradiometer, Bands 1 to 7, corrected for atmospheric conditions such as gases, aerosols, and Rayleigh scattering, at 500-m spatial resolution. In addition to the data, information on the quality of the reflectance bands can be obtained at 1000-m spatial resolution.

10.1.8 Enabling R packages

The R software is used as an example; however, other programs can be used to accomplish this task at the student's convenience.

The packages used in this homework assignment for determining complementary variables and mapping MODIS remote sensing data are `gdalUtils` (Greenberg, 2000), `raster`, `RStoolbox`, `rgdal`,

[1] https://earthexplorer.usgs.gov/
[2] https://ers.cr.usgs.gov/login/

`gridExtra` (Auguie & Antonov, 2017), `ggplot2` (Wickham et al., 2022), and `terra` (Hijmans et al., 2022). The packages are enabled with the `library` function in the R console.

```
library(gdalUtils)
library(raster)
library(RStoolbox)
library(rgdal)
library(gridExtra)
library(ggplot2)
library(terra)
```

10.1.9 Import files into R

The MOD09GA imagery product referring to Julian day 224, tile h12v10, from August 13, 2018, is obtained for analysis. The `get_subdatasets` function is used to access the `hdf` format files from R.

```
sds <- get_subdatasets("C:/sr/modis/MOD09GA.A2018224.h12v10.006.2018226034453.hdf")
sds
#[1] "HDF4_EOS:EOS_GRID:C:/sr/modis/
#MOD09GA.A2018224.h12v10.006.2018226034453.hdf:MODIS_Grid_1km_2D:num_observations_1km"
#[2] "HDF4_EOS:EOS_GRID:C:/sr/modis/
#MOD09GA.A2018224.h12v10.006.2018226034453.hdf:MODIS_Grid_1km_2D:state_1km_1"
#[3] "HDF4_EOS:EOS_GRID:C:/sr/modis/
#MOD09GA.A2018224.h12v10.006.2018226034453.hdf:MODIS_Grid_1km_2D:SensorZenith_1"
#[4] "HDF4_EOS:EOS_GRID:C:/sr/modis/
#MOD09GA.A2018224.h12v10.006.2018226034453.hdf:MODIS_Grid_1km_2D:SensorAzimuth_1"
#[5] "HDF4_EOS:EOS_GRID:C:/sr/modis/
#MOD09GA.A2018224.h12v10.006.2018226034453.hdf:MODIS_Grid_1km_2D:Range_1"
#[6] "HDF4_EOS:EOS_GRID:C:/sr/modis/
#MOD09GA.A2018224.h12v10.006.2018226034453.hdf:MODIS_Grid_1km_2D:SolarZenith_1"
#[7] "HDF4_EOS:EOS_GRID:C:/sr/modis/
#MOD09GA.A2018224.h12v10.006.2018226034453.hdf:MODIS_Grid_1km_2D:SolarAzimuth_1"
#[8] "HDF4_EOS:EOS_GRID:C:/sr/modis/
#MOD09GA.A2018224.h12v10.006.2018226034453.hdf:MODIS_Grid_1km_2D:gflags_1"
#[9] "HDF4_EOS:EOS_GRID:C:/sr/modis/
#MOD09GA.A2018224.h12v10.006.2018226034453.hdf:MODIS_Grid_1km_2D:orbit_pnt_1"
#[10] "HDF4_EOS:EOS_GRID:C:/sr/modis/
#MOD09GA.A2018224.h12v10.006.2018226034453.hdf:MODIS_Grid_1km_2D:granule_pnt_1"
#[11] "HDF4_EOS:EOS_GRID:C:/sr/modis/
#MOD09GA.A2018224.h12v10.006.2018226034453.hdf:MODIS_Grid_500m_2D:num_observations_500m"
#[12] "HDF4_EOS:EOS_GRID:C:/sr/modis/
#MOD09GA.A2018224.h12v10.006.2018226034453.hdf:MODIS_Grid_500m_2D:sur_refl_b01_1"
#[13] "HDF4_EOS:EOS_GRID:C:/sr/modis/
#MOD09GA.A2018224.h12v10.006.2018226034453.hdf:MODIS_Grid_500m_2D:sur_refl_b02_1"
#[14] "HDF4_EOS:EOS_GRID:C:/sr/modis/
#MOD09GA.A2018224.h12v10.006.2018226034453.hdf:MODIS_Grid_500m_2D:sur_refl_b03_1"
#[15] "HDF4_EOS:EOS_GRID:C:/sr/modis/
#MOD09GA.A2018224.h12v10.006.2018226034453.hdf:MODIS_Grid_500m_2D:sur_refl_b04_1"
#[16] "HDF4_EOS:EOS_GRID:C:/sr/modis/
#MOD09GA.A2018224.h12v10.006.2018226034453.hdf:MODIS_Grid_500m_2D:sur_refl_b05_1"
```

```
#[17] "HDF4_EOS:EOS_GRID:C:/sr/modis/
#MOD09GA.A2018224.h12v10.006.2018226034453.hdf:MODIS_Grid_500m_2D:sur_refl_b06_1"
#[18] "HDF4_EOS:EOS_GRID:C:/sr/modis/
#MOD09GA.A2018224.h12v10.006.2018226034453.hdf:MODIS_Grid_500m_2D:sur_refl_b07_1"
#[19] "HDF4_EOS:EOS_GRID:C:/sr/modis/
#MOD09GA.A2018224.h12v10.006.2018226034453.hdf:MODIS_Grid_500m_2D:QC_500m_1"
#[20] "HDF4_EOS:EOS_GRID:C:/sr/modis/
#MOD09GA.A2018224.h12v10.006.2018226034453.hdf:MODIS_Grid_500m_2D:obscov_500m_1"
#[21] "HDF4_EOS:EOS_GRID:C:/sr/modis/
#MOD09GA.A2018224.h12v10.006.2018226034453.hdf:MODIS_Grid_500m_2D:iobs_res_1"
#[22] "HDF4_EOS:EOS_GRID:C:/sr/modis/
#MOD09GA.A2018224.h12v10.006.2018226034453.hdf:MODIS_Grid_500m_2D:q_scan_1"
```

10.1.10 Convert file format

In this case, the main interest is to work only with reflectance bands 1 to 7, referring to items 12 to 18 listed in the `sds` object created earlier. The `gdal_translate` function is used to convert the data of interest into the `tif` format.

```
gdal_translate(sds[12], dst_dataset = "sur_refl_b01_1.tif")
gdal_translate(sds[13], dst_dataset = "sur_refl_b02_1.tif")
gdal_translate(sds[14], dst_dataset = "sur_refl_b03_1.tif")
gdal_translate(sds[15], dst_dataset = "sur_refl_b04_1.tif")
gdal_translate(sds[16], dst_dataset = "sur_refl_b05_1.tif")
gdal_translate(sds[17], dst_dataset = "sur_refl_b06_1.tif")
gdal_translate(sds[18], dst_dataset = "sur_refl_b07_1.tif")
```

The `raster` function is used to create seven objects of class `RasterLayer` for the seven reflectance bands, from the results of the `gdal_translate` function.

```
raster1 <- raster("sur_refl_b01_1.tif")
raster2 <- raster("sur_refl_b02_1.tif")
raster3 <- raster("sur_refl_b03_1.tif")
raster4 <- raster("sur_refl_b04_1.tif")
raster5 <- raster("sur_refl_b05_1.tif")
raster6 <- raster("sur_refl_b06_1.tif")
raster7 <- raster("sur_refl_b07_1.tif")
```

10.1.11 Transform map projection

Based on the file header information, you can check the sinusoidal map projection associated with the data, as in this band 1 header.

```
raster1
#class      : RasterLayer
#dimensions : 2400, 2400, 5760000  (nrow, ncol, ncell)
```

```
#resolution : 463.3127, 463.3127  (x, y)
#extent     : -6671703, -5559753, -2223901, -1111951  (xmin, xmax, ymin, ymax)
#crs        : +proj=sinu +lon_0=0 +x_0=0 +y_0=0 +R=6371007.181 +units=m +no_defs
#source     : sur_refl_b01_1.tif
#names      : sur_refl_b01_1
#values     : -32768, 32767  (min, max)
```

The `projectRaster` function is used to transform the sinusoidal projection to WGS-84.

```
sur_refl_b01_1prj <- projectRaster(
  raster1, crs=CRS('+proj=longlat +datum=WGS84 +no_defs +ellps=WGS84 +towgs84=0,0,0'))
sur_refl_b02_1prj <- projectRaster(
  raster2, crs=CRS('+proj=longlat +datum=WGS84 +no_defs +ellps=WGS84 +towgs84=0,0,0'))
sur_refl_b03_1prj <- projectRaster(
  raster3, crs=CRS('+proj=longlat +datum=WGS84 +no_defs +ellps=WGS84 +towgs84=0,0,0'))
sur_refl_b04_1prj <- projectRaster(
  raster4, crs=CRS('+proj=longlat +datum=WGS84 +no_defs +ellps=WGS84 +towgs84=0,0,0'))
sur_refl_b05_1prj <- projectRaster(
  raster5, crs=CRS('+proj=longlat +datum=WGS84 +no_defs +ellps=WGS84 +towgs84=0,0,0'))
sur_refl_b06_1prj <- projectRaster(
  raster6, crs=CRS('+proj=longlat +datum=WGS84 +no_defs +ellps=WGS84 +towgs84=0,0,0'))
sur_refl_b07_1prj <- projectRaster(
  raster7, crs=CRS('+proj=longlat +datum=WGS84 +no_defs +ellps=WGS84 +towgs84=0,0,0'))
```

The `terra` package can be used to perform the same function more efficiently. In this case, the data in class `RasterLayer` is transformed into class `SpatRaster` by using the `rast` function.

```
rast1 <- rast(raster1)
rast2 <- rast(raster2)
rast3 <- rast(raster3)
rast4 <- rast(raster4)
rast5 <- rast(raster5)
rast6 <- rast(raster6)
rast7 <- rast(raster7)
```

Then the projection transformation is performed with the `terra::project` function.

```
sur_refl_b01_1prj <- terra::project(
  rast1, '+proj=longlat +datum=WGS84 +no_defs +ellps=WGS84 +towgs84=0,0,0')
sur_refl_b02_1prj <- terra::project(
  rast2, '+proj=longlat +datum=WGS84 +no_defs +ellps=WGS84 +towgs84=0,0,0')
sur_refl_b03_1prj <- terra::project(
  rast3, '+proj=longlat +datum=WGS84 +no_defs +ellps=WGS84 +towgs84=0,0,0')
sur_refl_b04_1prj <- terra::project(
  rast4, '+proj=longlat +datum=WGS84 +no_defs +ellps=WGS84 +towgs84=0,0,0')
sur_refl_b05_1prj <- terra::project(
  rast5, '+proj=longlat +datum=WGS84 +no_defs +ellps=WGS84 +towgs84=0,0,0')
sur_refl_b06_1prj <- terra::project(
  rast6, '+proj=longlat +datum=WGS84 +no_defs +ellps=WGS84 +towgs84=0,0,0')
```

```
sur_refl_b07_1prj <- terra::project(
  rast7, '+proj=longlat +datum=WGS84 +no_defs +ellps=WGS84 +towgs84=0,0,0')
```

In the case of using the `projectRaster` function, the `RasterLayer` class data in the WGS-84 projection is stacked with the `stack` function to perform a first mapping of the MODIS h12v10 tile as a whole.

```
modis <- stack(sur_refl_b01_1prj, sur_refl_b02_1prj, sur_refl_b03_1prj,
  sur_refl_b04_1prj, sur_refl_b05_1prj, sur_refl_b06_1prj, sur_refl_b07_1prj)
```

In the case of using the `terra::project` function, the data is stacked just by using the atomic vector `c()`.

```
modis <- c(sur_refl_b01_1prj, sur_refl_b02_1prj, sur_refl_b03_1prj, sur_refl_b04_1prj,
             sur_refl_b05_1prj, sur_refl_b06_1prj, sur_refl_b07_1prj)
```

10.1.12 Renaming spectral bands

The stacked image data is renamed with the `names` function.

```
names(modis) <- c('red', 'NIR', 'blue', 'green', 'SWIR1', 'SWIR2', 'SWIR3')
modis
#class       : SpatRaster
#dimensions  : 2687, 3515, 7  (nrow, ncol, nlyr)
#resolution  : 0.003721507, 0.003721507  (x, y)
#extent      : -63.85067, -50.76957, -19.99969, -10  (xmin, xmax, ymin, ymax)
#coord. ref.: +proj=longlat +ellps=WGS84 +towgs84=0,0,0,0,0,0,0 +no_defs
#sources     : memory
#              memory
#              memory
#              ... and 4 more source(s)
#names       :        red,         NIR,        blue,       green,       SWIR1, SWIR2, ...
#min values  :-1000000.00,-730000.00,-1000000.00,-1000000.00,-1000000.00, -36065.52, ...
#max values  :   97681963,    98452637,    90517422,    96163008,    91184453, 67097310, ...
```

10.1.13 Stack spectral bands

The data stacked in the `SpatRaster` class is converted to the `RasterStack` class for mapping with functions from the `ggplot2` and `RStoolbox` packages.

```
modstack <- stack(modis)
```

10.1.14 Assign a map projection

With this procedure, the `crs` information was lost and it is necessary to reassign the `WGS-84` geodetic projection to the stacked images.

```
modstack <- "+proj=longlat +datum=WGS84 +no_defs +ellps=WGS84 +towgs84=0,0,0"
modstack
#class      : RasterStack
#dimensions: 2687, 3515, 9444805, 7  (nrow, ncol, ncell, nlayers)
#resolution: 0.003721507, 0.003721507  (x, y)
#extent     : -63.85067, -50.76957, -19.99969, -10  (xmin, xmax, ymin, ymax)
#crs        : +proj=longlat +ellps=WGS84 +towgs84=0,0,0,0,0,0,0 +no_defs
#names    : red,         NIR,         blue,        green,        SWIR1,        SWIR2,   SWIR3
#min values:-1000000.00,-730000.00,-1000000.00,-1000000.00,-1000000.00,-36065.52,-79816.13
#max values: 97681963,  98452637,  90517422,  96163008, 91184453,67097310,  63366167
```

10.1.15 Image previous mapping

A previous mapping of MODIS imagery data is performed in natural color and false-color infrared composition for visual inspection of the results.

```
a<-ggRGB(modstack, r = 1, g = 4, b = 3, stretch = "lin") + ggtitle("R1G4B3")
b<-ggRGB(modstack, r = 2, g = 4, b = 3, stretch = "lin") + ggtitle("R2G4B3")
grid.arrange(a,b, ncol=2)
```

The MODIS tile used covers almost the entire territorial extent of the state of Mato Grosso, with the scene located approximately in the center of the state. With the infrared composition it is possible to highlight regions of dense forest in the north of the state, and to the south, less dense, but representative in the Pantanal region. Regions where agriculture is practiced were also apparent in the central and eastern regions of the state in light brown tones (Figure 10.1).

10.1.16 Define geographic region and map subset of imagery data

The spatial polygon with attributes of the Upper and Middle Paraguay River basin is used to make a subset of the stacked MODIS imagery. The polygon is imported into R with the `readOGR` function.

```
baciasMT <- readOGR(dsn="C:/sr/modis/shp/baciasWGS84.shp", "baciasWGS84")
```

The `crop` and `mask` functions from the `raster` package are used to crop and mask pixels around the watershed.

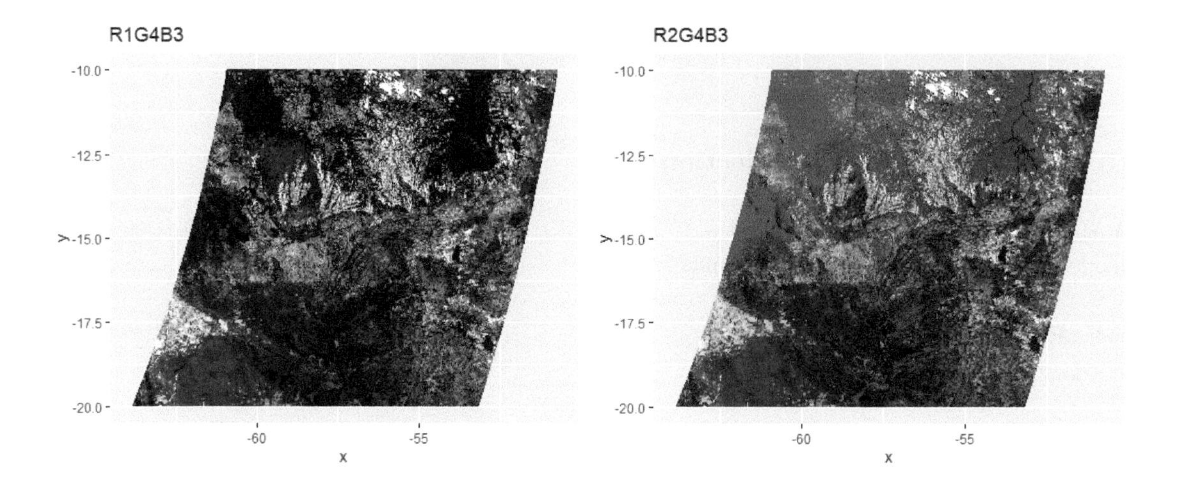

FIGURE 10.1 Natural color and false-color composition mapping from MODIS imagery related to August 13, 2018 in part of the state of Mato Grosso, Brazil.

```
modis_crop <- crop(modstack, baciasMT)
modis_mask <- mask(modis_crop, baciasMT)
```

MODIS image mapping in the watershed is performed in color compositions R1G4B3, R2G4B3, R7G6B2, and R7G6B5. The `ggRGB` function is used for mapping bands in RGB color channels with linear stretch.

```
a<-ggRGB(modis_mask, r = 1, g = 4, b = 3, stretch = "lin") + ggtitle("R1G4B3")
b<-ggRGB(modis_mask, r = 2, g = 4, b = 3, stretch = "lin") + ggtitle("R2G4B3")
c<-ggRGB(modis_mask, r = 7, g = 6, b = 2, stretch = "lin") + ggtitle("R7G6B2")
d<-ggRGB(modis_mask, r = 7, g = 6, b = 5, stretch = "lin") + ggtitle("R7G6B5")
grid.arrange(a,b,c,d, ncol=2)
```

With the compositions performed, the vegetation is well enhanced when infrared bands are used that clarify the real nature of the targets in the watershed when compared to the natural color composition, also making apparent locations with bare soil, water, water associated with riparian vegetation, urban areas, and crop and pasture regions (Figure 10.2).

10.1.17 Export results in `geotiff` format

The stacked, clipped and masked MODIS imagery bands around the watershed are converted to `SpatRaster` to be later exported by the `terra` package.

```
mod <- rast(modis_mask)
#mod
#class       : SpatRaster
#dimensions  : 565, 651, 7  (nrow, ncol, nlyr)
#resolution  : 0.003721507, 0.003721507  (x, y)
```

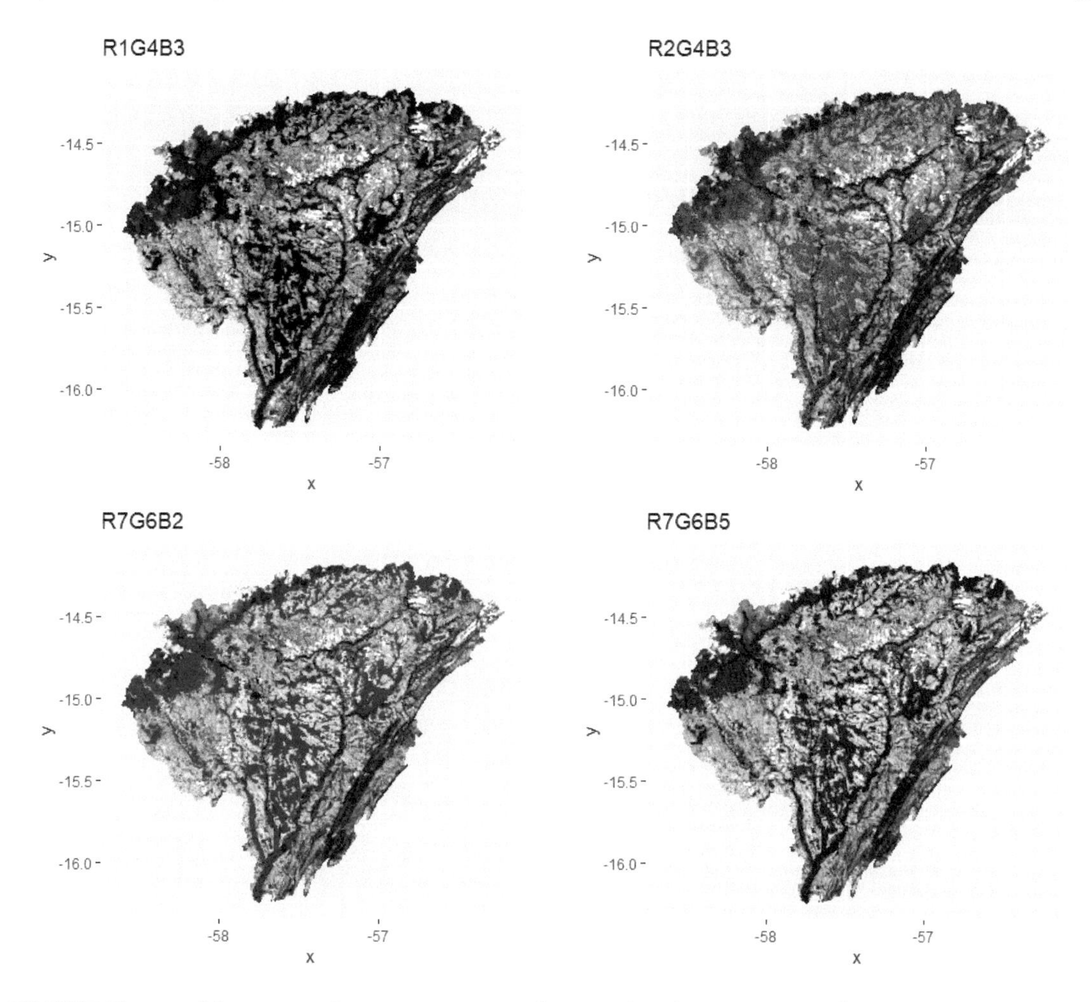

FIGURE 10.2 Mapping color compositions from MODIS imagery on August 13, 2018 in the Upper and Middle Paraguay watershed, state of Mato Grosso, Brazil.

```
#extent      : -58.63311, -56.21041, -16.24841, -14.14576  (xmin, xmax, ymin, ymax)
#coord. ref. : +proj=longlat +ellps=WGS84 +towgs84=0,0,0,0,0,0,0 +no_defs
#source      : memory
#names       :       red,       NIR,      blue,     green,     SWIR1,     SWIR2, ...
#min values  :  928501.9, 1651826.6,  588574.2, 1812443.8, 2769186.7, 1690946.0, ...
#max values  :  22110610,  41580029,   9930547,  14702689,  54204956,  51488838, ...
```

The `writeRaster` function is used to export a file in `geotiff` format with the bands stacked in the `SpatRaster` class into a directory of interest on the computer.

```
terra::writeRaster(mod, "C:/sr/modis/mod.tif", filetype = "GTiff", overwrite = TRUE)
```

10.1.18 Determining spectral indices

The exported file with stacked images can be imported back into R with the same export settings using the `stack` function.

```
mod <- stack("C:/sr/modis/mod.tif")
```

Afterwards, a set of big data can be obtained as a function of the pre-processed spectral bands. Spectral indices "CTVI", "DVI", "EVI", "EVI2", "GEMI", "GNDVI", "MNDWI", "MSAVI", "MSAVI2", "NBRI", "NDVI", "NDWI", "NDWI2", "NRVI", "RVI", "SATVI", "SAVI", "SLAVI", "SR", "TTVI", "TVI", "WDVI", are determined with the `spectralIndices` function.

```
si <- spectralIndices(mod, blue = "blue", green= "green", red = "red",
      nir = "NIR", swir2="SWIR2", swir3="SWIR3", scaleFactor = 100000000,
      coefs = list(L = 0.5, G = 2.5, L_evi = 1, C1 = 6, C2 = 7.5, s = 1))
```

10.1.19 Determining tasseled cap bands

Additionally, the tasseled cap transformation can be used to determine brightness, greenness, and wetness bands from MODIS imagery (Lobser & Cohen, 2007).

The `tasseledCap` function is used with the calculation settings for MODIS to obtain three new bands.

```
mod_tc <- tasseledCap(mod, sat = "MODIS")
```

10.1.20 Mapping tasseled cap results

The results of calculated spectral indices and the `tasseledCap` transformation are stacked in a single file with the `stack` function.

```
modbd <- stack(si, mod_tc)
```

The total data set obtained is mapped with the `plot` function (Figure 10.3).

```
plot(modbd, maxnl = 25)
```

The `ggRGB` function is used to map a color composition with the brightness, greenness, and wetness bands in the R, G, and B channels, respectively.

```
ggRGB(modbd, r = 23, g = 24, b = 25, stretch = "lin") + ggtitle("RbGgBw")
```

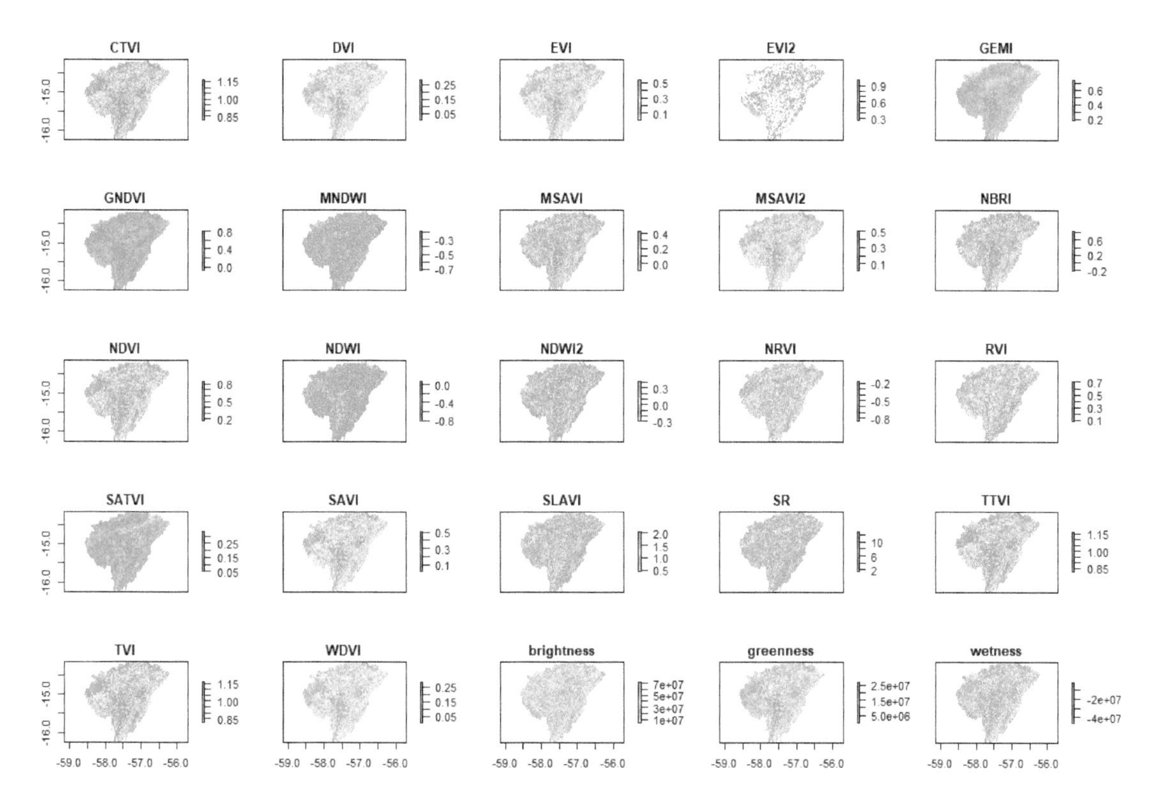

FIGURE 10.3 Mapping spectral indices and tasseled cap components from MODIS imagery on August 13, 2018 in the Upper and Middle Paraguay watershed, state of Mato Grosso, Brazil.

Thereby, it can be seen that regions with water, dense natural vegetation and lithological geoforms in a parallel pattern to the south of the image are mapped in blue tones in the color composition, due to higher apparent humidity in these targets. The remaining urban areas, bare soil and grassland vegetation are mapped in red tones (Figure 10.4).

The relationship between the brightness, greenness, and wetness bands can be graphically represented by a scatterplot of the possible combinations of bands. The `plot` function is used to make the plots with the `$` operator selecting each variable of interest.

```
par(mfrow = c(2,2))
plot(modbd$brightness, modbd$greenness)
plot(modbd$brightness, modbd$wetness)
plot(modbd$greenness, modbd$wetness)
```

With this it is possible to verify an inverse relationship between wetness and brightness and a positive relationship between wetness and greenness (Figure 10.5).

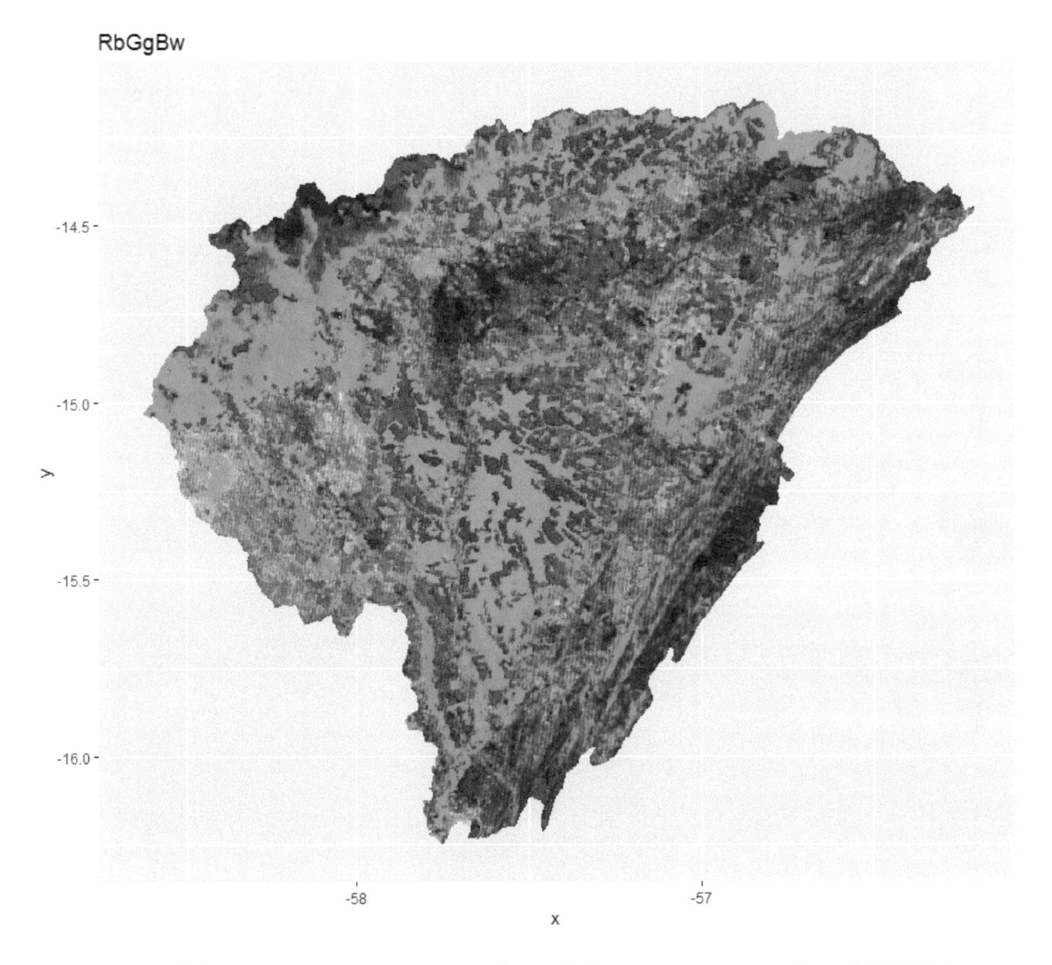

FIGURE 10.4 Color composition mapping of tasseled cap components from MODIS imagery on August 13, 2018 in the Upper and Middle Paraguay watershed, state of Mato Grosso, Brazil.

10.2 Solved Exercises

10.2.1 In additive color theory, a pixel with RGB digital values of 255,255,255 will show white color. In the case of using 3 images of 8 bits in the additive theory, how many different digital numbers in the RGB coordinate system are represented?

```
nd<-2^24
nd
```

```
## [1] 16777216
```

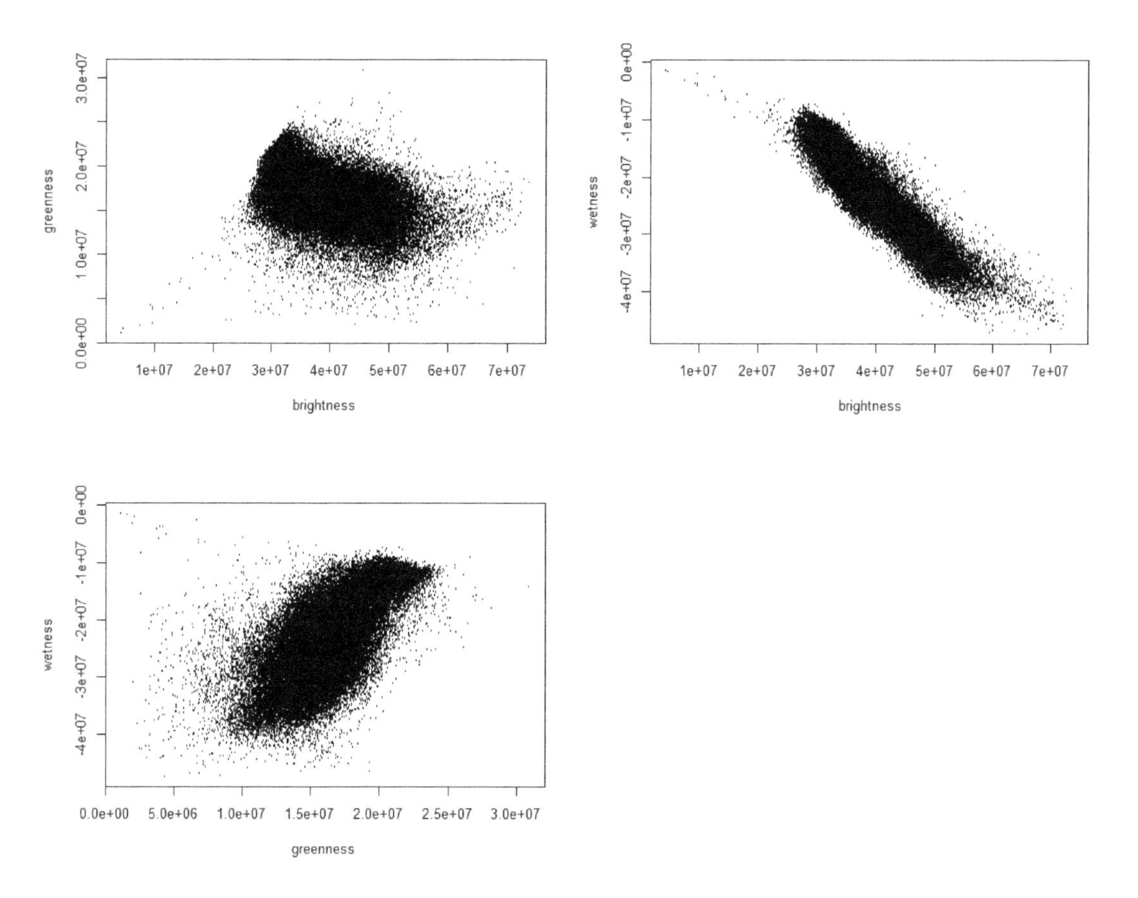

FIGURE 10.5 Scatterplot performed between tasseled cap bands from MODIS imagery on August 13, 2018 in the Upper and Middle Paraguay watershed, state of Mato Grosso, Brazil.

10.2.2 A linear contrast is applied to a Landsat-5 image with 8 bits of radiometric resolution. The minimum and maximum brightness values are 4 and 105, respectively. Determine the output brightness values for the original minimum and maximum brightness values.

```
# BVs=[(BVe-mink)/(maxk-mink)]quantk
BVsmin <- ((4-4)/(105-4))*255
BVsmin
```

```
## [1] 0
```

```
BVsmax <- ((105-4)/(105-4))*255
BVsmax
```

```
## [1] 255
```

A: The output brightness values for the original minimum and maximum brightness values are 0 and 255, respectively.

10.2.3 Define “image interpretation”.

A: Image interpretation is defined as the examination of images to identify objects and assess their meaning.

10.2.4 List the aspects that make image interpretation a scientific tool.

A: Aerial and regional perspective, three-dimensional depth perception, ability to gain knowledge beyond human visual perception, and ability to obtain historical images to detect change.

10.2.5 Explain how research methods are applied to image interpretation of higher orders of complexity.

A: Scientific approaches can be used in interpreting remote sensing data based on complementary information, evidence convergence, and multi-concept in image analysis.

10.2.6 Define color and false-color compositions used in the practical work justifying the choice.

A: Some color and false-color compositions with linear enhancement can be chosen according to the object of interest being monitored by the Landsat-8 satellite (Table 10.1):

TABLE 10.1 Suggested target types and color compositions based on Landsat-8 OLI spectral bands.

Target	RGB Composition
Agriculture	6,5,4
Vegetation	6,5,4
Infrared vegetation	5,4,3
Land and water	5,6,4
Urban false-color	7,6,4
Infrared	7,5,4
Atmospheric penetration	7,6,5

10.2.7 Cite four graphical representation commands used in processing remotely sensed digital images in R.

R: `plot()`, `plotRGB()`, `ggR()`, `ggRGB()`.

10.2.8 Justify the use of a false-color composition to detect vegetation in municipalities in southern Minas Gerais state.

A: The R5G4B3 composition with linear enhancement can be used to make the vegetation appear in red tones and be visually distinct from the others.

11

Unsupervised Classification of Remote Sensing Images

11.1 Homework Solution

11.1.1 Subject

Unsupervised segmentation analysis applied to multispectral imaging from the CBERS-04A satellite monitoring of urban and rural environments.

11.1.2 Abstract

Superpixels group visually similar pixels together to create visually meaningful entities and reduce the number of primitive elements in subsequent processing steps. We aimed to obtain remote sensing data and perform unsupervised segmentation for pattern recognition of multispectral imaging from the CBERS-04A satellite, WPM camera, from August 15, 2021. Superpixel segmentation is used to propose a classification scheme for landscape targets from normalized difference vegetation index (NDVI) and RGB bands with and without panchromatic band fusion.

Keywords: Image fusion, monitoring, NDVI, stretch, SLIC, WPM camera.

11.1.3 Introduction

Remote sensing imagery needs to be converted into tangible information that can be used together with other datasets in geoprocessing. Around the year 2000, geoprocessing and image processing began to grow rapidly together through object-based image analysis (OBIA) (Blaschke, 2010). Thus, methods have been developed for application in remote sensing by establishing segmentation procedures and developing software with improvements to detect objects with greater accuracy (Esch et al., 2008). R packages have also been developed for pattern-based spatial analysis with support for multi-layer raster datasets for landscape target classification (Nowosad, 2021).

Superpixels group visually similar pixels together to create visually meaningful entities, while reducing the number of primitive elements for subsequent processing steps (Stutz et al., 2018). With the use of superpixels in computer vision algorithms, it has become possible to capture image redundancy, create geometries from computations performed on the image, and reduce the complexity of image processing tasks (Achanta et al., 2011). Considering the need to recognize patterns in images for data classification, superpixels are a promising group of techniques that allow generalization of spatial information. Among this group, the Simple Linear Iterative Clustering (SLIC) algorithm has proven to be of high quality and performance (Achanta et al., 2011; Nowosad & Stepinski, 2021). Superpixels refer to segmentation concepts used to group pixels with similar characteristics. The concept of superpixels can be applied in computer vision both in single-layer images and to delineate segments in color composition of RGB images. In both cases, the methods are used as approaches to classify more significant regions in the image associated with simplifying results to facilitate pattern detection in the image and target recognition. When the superpixel technique is applied to RGB images, each superpixel resulting from the analysis contains similar colors that can be used to reliably represent objects. SLIC superpixels can be used for fast and accurate thematic mapping to improve the automation of analysis, processing and labeling of large remote sensing data, especially when time is a constraining factor (Csillik, 2017).

The `supercells` package (Nowosad et al., 2022) uses the concept of superpixels for a variety of spatial data, in a continuous normalized vegetation index raster or multiple variables, as in an RGB raster. Spatial patterns can also be detected in categorical raster. Results from areas with similar characteristics are regionalized and it is possible to compare values of one or more variables (Nowosad & Stepinski, 2021).

11.1.4 Objective

The objective of the homework is to obtain remote sensing data and perform unsupervised segmentation for pattern recognition of multispectral imagery from the CBERS-04A satellite, WPM camera, from August 15, 2021. It is intended to detect pixel patterns by segmentation and propose a classification scheme for landscape targets from the normalized difference vegetation index (NDVI) determined from CBERS-04A imagery. Furthermore, a segmentation approach is applied to RGB bands to simplify regions with similar targets and recognize R, G, and B values in these regions. RGB indices can be determined for each superpixel to form a database of the region.

11.1.5 Acquiring images by web browsing

CBERS-04A monitoring data from the WPM camera are used as an example of panchromatic band mapping at 2-m spatial resolution in monitoring performed on August 15, 2021 at the Funil dam. The data obtained refer to the L4 level, corresponding to the orthorectified image with radiometric and geometric correction using control points and a digital elevation model.

11.1.6 Enabling R packages

The R packages `supercells` (Nowosad et al., 2022), `terra` (Hijmans et al., 2022), `sf` (Pebesma et al., 2021), `dplyr` (Wickham et al., 2021), `ggplot2` (Wickham et al., 2022), and `gridExtra` (Auguie & Antonov, 2017) are used and enabled for use with the `library` function.

```
library(supercells)
library(terra)
library(sf)
library(dplyr)
library(ggplot2)
library(gridExtra)
```

A single file of the Brovey fused bands is created by adding the fused NIR band to the dataset with the fused RGB bands using the `addLayer` function. The non-fused multispectral bands are resampled to the same resolution of the panchromatic band by the nearest neighbor method to enable stacking the fused and non-fused images in a single file. The nearest neighbor method is used to provide less modification to the original data by improving the spatial resolution of the raster from 8 m to 2 m with this type of interpolation. The rasters are transformed into the `SpatRaster` class to perform this operation with the `rast` function. The resampled data is converted back to the `raster` class to generate a single data file with both fused and non-fused images. The file ready for data analysis is imported into R with the `rast` function.

```
c4rast <- rast("C:/sr/c11/c4rast1.tif")
c4rast
#class       : SpatRaster
#dimensions  : 680, 772, 9  (nrow, ncol, nlyr)
#resolution  : 2, 2  (x, y)
#extent      : 501788, 503332, 7652926, 7654286  (xmin, xmax, ymin, ymax)
#coord. ref. : WGS 84 / UTM zone 23S (EPSG:32723)
#source      : c4rast1.tif
#names       :      blue,     green,       red,       nir, blue_pan, green_pan, ...
#min values  : 156.00000,  95.00000,  87.00000,  74.00000, 40.14257,  29.32076, ...
#max values  :  641.0000,  619.0000,  730.0000,  699.0000, 301.3992,  327.3260, ...
```

Based on the raster file header, it is possible to see the order of bands that will be used in the comparative color geovisualization between multispectral and fused bands. The `par(mfrow)` function is used to map the images side by side and the `plotRGB` function to map the color compositions.

```
par(mfrow=c(1,2))
plotRGB(c4rast, r=3, g=2, b=1, stretch="lin")
plotRGB(c4rast, r=7, g=6, b=5, stretch="lin")
```

Thus, it is possible to observe greater sharpness of imagery details from the fused bands, despite the same spatial resolution of the images. However, the colors of the fused bands are modified in relation to the multispectral bands and become psychedelic tones (Figure 11.1).

FIGURE 11.1 Color composition mapping of multispectral data (left) and multispectral fusion with the panchromatic band (right) from the CBERS-04A imagery, sensor WPM, in the region of Lavras, state of Minas Gerais, Brazil, on August 15, 2021.

A function is created to calculate the normalized difference vegetation index (NDVI) from the near-infrared and red spectral bands.

```
ndvi_fun <- function(nir, red){(nir - red) / (nir + red)}
```

With this, it is possible to verify that the NDVI result is variable according to the input data, showing greater magnitude of variation when obtained from multispectral bands (-0.31 to 0.61) compared to fused bands (-0.23 to 0.41).

```
# Multispectral
ndvi <- lapp(c4rast[[c(4, 3)]], fun = ndvi_fun)
ndvi
#class       : SpatRaster
#dimensions  : 680, 772, 1  (nrow, ncol, nlyr)
#resolution  : 2, 2  (x, y)
#extent      : 501788, 503332, 7652926, 7654286  (xmin, xmax, ymin, ymax)
#coord. ref. : WGS 84 / UTM zone 23S (EPSG:32723)
#source      : memory
#name        :       lyr1
#min value   : -0.3178947
#max value   :  0.6113903

# Fusion
ndvip <- lapp(c4rast[[c(8, 7)]], fun = ndvi_fun)
ndvip
#class       : SpatRaster
#dimensions  : 680, 772, 1  (nrow, ncol, nlyr)
#resolution  : 2, 2  (x, y)
#extent      : 501788, 503332, 7652926, 7654286  (xmin, xmax, ymin, ymax)
#coord. ref. : WGS 84 / UTM zone 23S (EPSG:32723)
#source      : memory
#name        :       lyr1
#min value   : -0.2368906
#max value   :  0.4172197
```

The superpixel segmentation is applied to the NDVI data with the `supercells` function. In both cases the values of k and compactness are 10000 and 10, respectively.

```
# Multispectral
ndvisc <- supercells(ndvi, k = 10000, compactness = 10)
ndvisc
#Simple feature collection with 8160 features and 4 fields
#Geometry type: MULTIPOLYGON
#Dimension:     XY
#Bounding box:  xmin: 501788 ymin: 7652926 xmax: 503332 ymax: 7654286
#Projected CRS: WGS 84 / UTM zone 23S
#First 10 features:
#   supercells      x       y      lyr1                       geometry
#1           1 501795 7654279 0.1402409 MULTIPOLYGON (((501788 7654...
#2           2 501793 7654263 0.1486910 MULTIPOLYGON (((501788 7654...
#3           3 501793 7654249 0.1441218 MULTIPOLYGON (((501788 7654...
#4           4 501795 7654233 0.1360560 MULTIPOLYGON (((501794 7654...
#5           5 501791 7654215 0.1432107 MULTIPOLYGON (((501788 7654...
#6           6 501795 7654199 0.1543498 MULTIPOLYGON (((501794 7654...
```

```
#7          7 501795 7654183 0.1446506 MULTIPOLYGON (((501788 7654...
#8          8 501795 7654167 0.1537297 MULTIPOLYGON (((501788 7654...
#9          9 501793 7654153 0.1744087 MULTIPOLYGON (((501788 7654...
#10        10 501793 7654139 0.1507382 MULTIPOLYGON (((501788 7654...

# Fusion
ndviscp <- supercells(ndvip, k = 10000, compactness = 10)
ndviscp
#Simple feature collection with 8160 features and 4 fields
#Geometry type: MULTIPOLYGON
#Dimension:     XY
#Bounding box:  xmin: 501788 ymin: 7652926 xmax: 503332 ymax: 7654286
#Projected CRS: WGS 84 / UTM zone 23S
#First 10 features:
#   supercells      x       y       lyr1                       geometry
#1           1 501795 7654279 0.08328568 MULTIPOLYGON (((501788 7654...
#2           2 501795 7654263 0.08948024 MULTIPOLYGON (((501788 7654...
#3           3 501795 7654247 0.08520211 MULTIPOLYGON (((501788 7654...
#4           4 501793 7654231 0.06702760 MULTIPOLYGON (((501788 7654...
#5           5 501793 7654217 0.08657206 MULTIPOLYGON (((501788 7654...
#6           6 501795 7654203 0.09531742 MULTIPOLYGON (((501794 7654...
#7           7 501793 7654185 0.08770134 MULTIPOLYGON (((501788 7654...
#8           8 501795 7654167 0.09176530 MULTIPOLYGON (((501788 7654...
#9           9 501795 7654151 0.10125023 MULTIPOLYGON (((501788 7654...
#10         10 501793 7654135 0.06313175 MULTIPOLYGON (((501788 7654...
```

Considering that the NDVI presented different values according to the input variables, different thresholds were set to characterize the superpixel variation in 5 user-defined classes. The `mutate` function is used to set the separate NDVI thresholds in each class.

```
# Multispectral
ndviclass <- ndvisc %>% mutate(class = case_when(lyr1 <= -0.12 ~ "1",
      lyr1 <= 0.18 ~ "2", lyr1 <= 0.21 ~ "3",  lyr1 <= 0.32 ~ "4",
                                              lyr1 > 0.32 ~ "5"))
ndviclass
#Simple feature collection with 8160 features and 5 fields
#Geometry type: MULTIPOLYGON
#Dimension:     XY
#Bounding box:  xmin: 501788 ymin: 7652926 xmax: 503332 ymax: 7654286
#Projected CRS: WGS 84 / UTM zone 23S
#First 10 features:
#   supercells      x       y      lyr1                       geometry class
#1           1 501795 7654279 0.1402409 MULTIPOLYGON (((501788 7654...     2
#2           2 501793 7654263 0.1486910 MULTIPOLYGON (((501788 7654...     2
#3           3 501793 7654249 0.1441218 MULTIPOLYGON (((501788 7654...     2
#4           4 501795 7654233 0.1360560 MULTIPOLYGON (((501794 7654...     2
#5           5 501791 7654215 0.1432107 MULTIPOLYGON (((501788 7654...     2
#6           6 501795 7654199 0.1543498 MULTIPOLYGON (((501794 7654...     2
#7           7 501795 7654183 0.1446506 MULTIPOLYGON (((501788 7654...     2
#8           8 501795 7654167 0.1537297 MULTIPOLYGON (((501788 7654...     2
#9           9 501793 7654153 0.1744087 MULTIPOLYGON (((501788 7654...     2
```

```
#10         10 501793 7654139 0.1507382 MULTIPOLYGON (((501788 7654...      2

ndviclassp <- ndviscp %>% mutate(class = case_when(lyr1 <= -0.05 ~ "1",
      lyr1 <= 0.11 ~ "2", lyr1 <= 0.14 ~ "3",  lyr1 <= 0.23 ~ "4",
                                                 lyr1 > 0.23 ~ "5"))
# Fusion
ndviclassp
#Simple feature collection with 8160 features and 5 fields
#Geometry type: MULTIPOLYGON
#Dimension:     XY
#Bounding box:  xmin: 501788 ymin: 7652926 xmax: 503332 ymax: 7654286
#Projected CRS: WGS 84 / UTM zone 23S
#First 10 features:
#   supercells      x       y       lyr1                      geometry class
#1           1 501795 7654279 0.08328568 MULTIPOLYGON (((501788 7654...     2
#2           2 501795 7654263 0.08948024 MULTIPOLYGON (((501788 7654...     2
#3           3 501795 7654247 0.08520211 MULTIPOLYGON (((501788 7654...     2
#4           4 501793 7654231 0.06702760 MULTIPOLYGON (((501788 7654...     2
#5           5 501793 7654217 0.08657206 MULTIPOLYGON (((501788 7654...     2
#6           6 501795 7654203 0.09531742 MULTIPOLYGON (((501794 7654...     2
#7           7 501793 7654185 0.08770134 MULTIPOLYGON (((501788 7654...     2
#8           8 501795 7654167 0.09176530 MULTIPOLYGON (((501788 7654...     2
#9           9 501795 7654151 0.10125023 MULTIPOLYGON (((501788 7654...     2
#10         10 501793 7654135 0.06313175 MULTIPOLYGON (((501788 7654...     2
```

The `ggplot` function is used to map the results. The `grid.arrange` function enables comparison of the mapping results side by side. The `scale_fill_discrete` function is used to replace the class numbers with the class name, in this case "Water", "Urban", "Bare soil", "Vegetation", and "Dense vegetation".

```
a<-ggplot() + geom_sf(data = ndviclass[,6], aes(fill = class)) +
   scale_fill_discrete(labels=c("Water", "Urban","Bare soil",
                              "Vegetation","Dense vegetation"),
    type=c("blue","grey90","brown3", "green","darkgreen"))
b<-ggplot() + geom_sf(data = ndviclassp[,6], aes(fill = class)) +
   scale_fill_discrete(labels=c("Water", "Urban","Bare soil",
                              "Vegetation","Dense vegetation"),
    type=c("blue","grey90","brown3", "green","darkgreen"))
grid.arrange(a, b, ncol=1)
```

According to the NDVI mapping thresholds, the results with and without the application of fusion to the multispectral bands showed little apparent difference, as can be observed in the maps (Figure 11.2). However, if the method is applied at the original spatial resolution of the CBERS-04A bands there may be greater difference when considering the use of the panchromatic band in the results. Nevertheless, based on the results, it is possible to establish an unsupervised classification methodology to remote sensing imagery data using vegetation vigor magnitude concepts in the classification of targets occurring in the landscape.

The results are exported into the directory of interest with the `st_write` function.

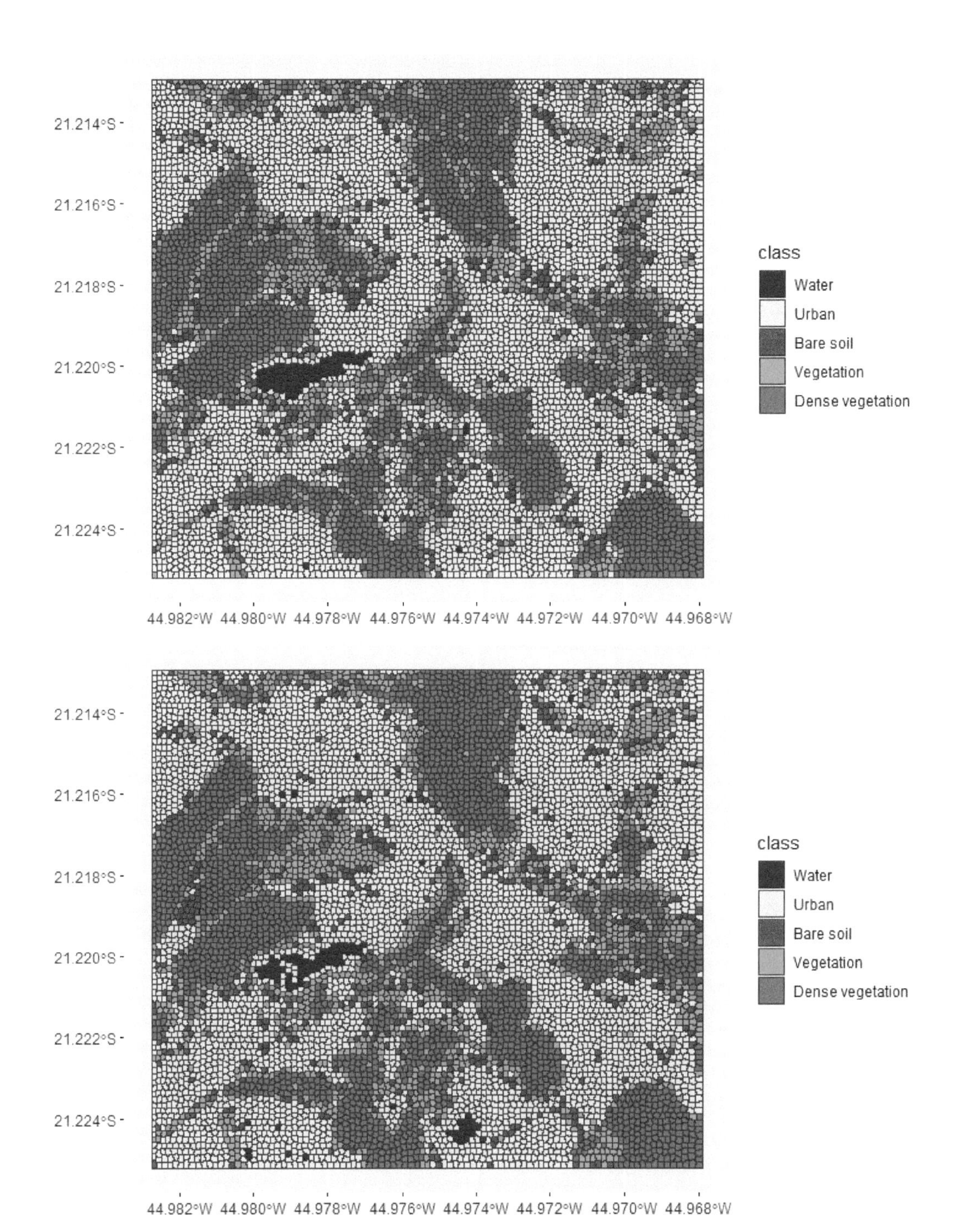

FIGURE 11.2 Supercells approach for unsupervised landscape classification based on thresholds established from NDVI calculated with data from the CBERS-04A satellite, WPM sensor, in the region of Lavras, state of Minas Gerais, Brazil, on August 15, 2021.

```
st_write(ndviclass, "C:/sr/c11/ndviclass.shp", delete_layer = TRUE)
st_write(ndviclassp, "C:/sr/c11/ndviclassp.shp", delete_layer = TRUE)
```

11.1.7 RGB image segmentation with superpixels

The superpixel technique can also be applied to RGB images. A subset of data is defined with the multispectral bands obtained in the spectral regions of blue, green and red to perform the segmentation. The `subset` function is used to define the subsets. The same image as in the previous example is used here.

```
c4 <- subset(c4rast, c(1,2,3))
```

A linear stretch is applied to the subset of data for better visualization of targets in the images. The `stretch` function is used with the `minv`, `maxv`, `minq` and `maxq` parameters to apply the enhancement to the desired configuration and to transform the radiometric data to 8 bits (0 to 255).

```
c4s <- stretch(c4, minv=0, maxv=255, minq=0.01, maxq=0.98)
c4s
#class       : SpatRaster
#dimensions  : 680, 772, 3  (nrow, ncol, nlyr)
#resolution  : 2, 2  (x, y)
#extent      : 501788, 503332, 7652926, 7654286  (xmin, xmax, ymin, ymax)
#coord. ref. : WGS 84 / UTM zone 23S (EPSG:32723)
#source      : memory
#names       : blue, green, red
#min values  :    0,     0,   0
#max values  :  255,   255, 255
```

The RGB function is used to create RGB images from the subsets of data.

```
RGB(c4s) <- c(3,2,1)
c4s
#class       : SpatRaster
#dimensions  : 680, 772, 3  (nrow, ncol, nlyr)
#resolution  : 2, 2  (x, y)
#extent      : 501788, 503332, 7652926, 7654286  (xmin, xmax, ymin, ymax)
#coord. ref. : WGS 84 / UTM zone 23S (EPSG:32723)
#source      : memory
#colors RGB  : 3, 2, 1
#names       : blue, green, red
#min values  :    0,     0,   0
#max values  :  255,   255, 255
#RGB(c4ps) <- c(3,2,1)
```

The `supercells` function is used to delimit areas with similar colors that can potentially represent the same objects in the RGB color composition. For an RGB image, an additional argument called "transform ="to_LAB" " is used to perform an internal calculation in the LAB color space instead of RGB.

```
slic1 <- supercells(c4s, k = 10000, compactness = 10, transform = "to_LAB")
slic1
#Simple feature collection with 7809 features and 6 fields
#Geometry type: MULTIPOLYGON
#Dimension:     XY
#Bounding box:  xmin: 501788 ymin: 7652926 xmax: 503332 ymax: 7654286
#Projected CRS: WGS 84 / UTM zone 23S
#First 10 features:
#   supercells      x       y      blue     green      red                       geometry
#1   1 501793 7654279   93.01635  99.44745 109.4052 MULTIPOLYGON (((501788 7654...
#2   2 501795 7654263   93.67170  98.25660 108.0358 MULTIPOLYGON (((501796 7654...
#3   3 501795 7654247 102.49215 107.95425 115.7190 MULTIPOLYGON (((501788 7654...
#4   4 501795 7654231   99.79680 112.61565 125.4268 MULTIPOLYGON (((501788 7654...
#5   5 501795 7654215   97.54005  98.62635 106.3070 MULTIPOLYGON (((501788 7654...
#6   6 501795 7654199   97.21875  99.88860 106.9572 MULTIPOLYGON (((501788 7654...
#7   7 501795 7654181   94.66875  99.20010 108.7116 MULTIPOLYGON (((501788 7654...
#8   8 501795 7654163   92.08305  96.57870 107.8599 MULTIPOLYGON (((501788 7654...
#9   9 501797 7654143 114.54090 134.46660 138.2074 MULTIPOLYGON (((501788 7654...
#10 10 501795 7654131 103.82070 104.54490 135.1704 MULTIPOLYGON (((501788 7654...
```

The `slic1` output is an `sf` object, where each row stores the id of each superpixel, x, y coordinates of superpixel centroids, and an average of all input variables in red, green and blue colors. A first visualization of results is performed by overlaying superpixels on top of the original RGB image with the `plot` function (Figure 11.3).

```
plot(c4s)
plot(st_geometry(slic1), add = TRUE, border="grey")
```

The second type of visualization first requires converting the average RGB values into hexadecimal representation. A function is defined to perform this conversion.

```
rgb_to_hex <- function(x){
  apply(t(x), 2, function(x) rgb(x[1], x[2], x[3], maxColorValue = 255))
}
```

The `rgb_to_hex` function defined earlier is applied to the output file `slic1` obtained by applying the `supercells` function.

```
avg_colors1 <- rgb_to_hex(st_drop_geometry(slic1[4:6]))
```

The hexadecimal color results are mapped onto each superpixel without defining border lines to the vectors.

```
plot(st_geometry(slic1), border = NA, col = avg_colors1)
```

You can notice that the visualization obtained is not an image but a set of colored superpixels (Figure 11.3).

FIGURE 11.3 Mapping of superpixels defined from the RGB image obtained from the CBERS-04A satellite monitoring, WPM sensor, in the Lavras region, state of Minas Gerais, Brazil, on August 15, 2021.

Therefore, instead of representing the region by 524960 cells ($(680 * 772)^3$), 7809 features and 4 fields are used. From this, a data labeling can be proposed in a classification scheme to be defined by the analyst.

11.2 Solved Exercises

11.2.1 Briefly define digital image classification.

A: Digital image classification is the process of assigning classes to pixels.

FIGURE 11.4 Hexadecimal color mapping in superpixels defined from RGB image obtained from CBERS-04A satellite monitoring, sensor WPM, in the region of Lavras, state of Minas Gerais, Brazil, on August 15, 2021.

11.2.2 What are the advantages and disadvantages of unsupervised classification of digital images?

A: The advantages of unsupervised classification when compared to supervised classification are: No prior knowledge of the region to perform classification; human error is minimized; unique classes are recognized as distinct units.

The disadvantages of unsupervised classification when compared to supervised classification are: The classifier may recognize homogeneous spectral classes that do not match the analyst's categories of interest; the analyst has little control over existing classes and specific identity; classes may change over time according to the spectral variation of the monitored objects, determining different classification results of the same previously classified target.

11.2.3 Cite three methods of defining distances in attribute space for unsupervised classification of digital images.

A: Euclidean distance, minimum distance, and k-medoids.

11.2.4 Cite three algorithms for unsupervised classification of digital images.

A: k-means, CLARA, and k-means random forest.

11.2.5 What is the basic difference between pixel-by-pixel and object-based classification?

A: In pixel-by-pixel classification, the smallest unit of information used in the classification is one pixel from each spectral band used. In object-based classification, a group of pixels with homogeneous color defines a segmented region in the classified image.

11.2.6 Determine the Euclidean distance of pixels.

Pixel values are presented to determine the Euclidean distance between vertices A and B in bands 1 to 4 of Landsat-8 image (Table 11.1).

TABLE 11.1 Values of pixels A and B in multiple bands of Landsat-8.

Pixel	Band 1	Band 2	Band 3	Band 4
A	34	28	22	6
B	26	16	52	29

A: The distance AB is 40.45 (Table 11.2).

TABLE 11.2 Determination of Euclidean distance between pixels A and B in multiple bands of Landsat-8.

Calculation	Band 1	Band 2	Band 3	Band 4
$Difference\ A, B$	8	12	-30	-23
$(Difference\ A, B)^2$	64	144	900	529
$\sqrt{\sum(Difference\ A, B)^2}$	40.45	—	—	—

Computational determination:

```
difAB<-sum((34-26)^2, (28-16)^2, (22-52)^2, (6-29)^2)
distAB<-sqrt(difAB)
distAB
```

```
## [1] 40.45986
```

11.2.7 Classifying pixel data A, B, and C

Given that the interest of a remote sensing analyst is to classify pixel data A, B, and C, determine the Euclidean distance between A, B; B, C and A, C and propose a criterion for separating the pixels into only two classes (Table 11.3).

TABLE 11.3 Values of pixels A, B, and C in multiple bands of Landsat-8.

Pixel	Band 1	Band 2	Band 3	Band 4
A	5	10	15	15
B	25	30	35	38
C	18	25	23	20

Computational determination:

```
difAB<-sum((5-25)^2, (10-30)^2, (15-35)^2, (15-38)^2)
distAB<-sqrt(difAB)
distAB
```

```
## [1] 41.58125
```

```
difBC<-sum((25-18)^2, (30-25)^2, (35-23)^2, (38-20)^2)
distBC<-sqrt(difBC)
distBC
```

```
## [1] 23.28089
```

```
difAC<-sum((5-18)^2, (10-25)^2, (15-23)^2, (15-20)^2)
distAC<-sqrt(difAC)
distAC
```

```
## [1] 21.97726
```

The distance AC is the smallest distance, followed by BC and AB. Therefore, the two classes are composed of the pixels AC and B, respectively.

12

Supervised Classification of Remote Sensing Images

12.1 Homework Solution

12.1.1 Subject

Supervised classification analysis applied to multispectral imaging from the CBERS-04A satellite for environmental monitoring.

12.1.2 Abstract

Supervised learning used in image classification enables labeling image cells and evaluating the quality of predictions from input data sets. We aimed to obtain remote sensing data and perform supervised classification for pattern recognition of multispectral imagery from the CBERS-04A satellite, WPM camera, August 15, 2021. Spatial polygons are created under different arrangements of 10 and 4 classes to evaluate the effect of class generalization on training supervised classification algorithms on CBERS-04A multispectral imaging data with and without panchromatic band fusion. Exploratory analysis of training data is performed to evaluate spectral characteristics of samples obtained from the monitored targets. Importance of variables and decision trees are created to evaluate the decision-making reasoning of algorithms in classification. With visual analysis of mapping results, it is difficult to evaluate differences that determine the choice of the best classification algorithm, which can be better determined after accuracy analysis.

Keywords: Classification algorithms, exploratory analysis, image fusion, WPM camera, machine learning, pattern detection.

12.1.3 Introduction

Supervised classification is a frequently used procedure for quantitative analysis of remote sensing image data. Algorithms are used in this process to recognize and label pixels in an image to represent specific types of land use and landscape analysis. Supervised learning is a branch of machine learning, a data analysis method that uses algorithms that iteratively learn from data to find information in images after model fitting and spatial prediction of classes in the input data provided to the system (Campbell & Wynne, 2011; Jensen, 2005; Kuhn et al., 2020).

12.1.4 Objective

The objective of the homework is to create spatial polygons under different class arrangements and evaluate the training of supervised classification algorithms in spatial class prediction on CBERS-04A multispectral imaging data with and without panchromatic band fusion. In addition, an exploratory analysis of training data is performed to evaluate spectral characteristics of the samples obtained from the monitored targets.

12.1.5 CBERS-04A data specifications

As an example performed by the teacher, CBERS-04A data is used; however, a choice of other data can be made at the student's discretion. The WPM camera is the main payload of CBERS 04A, providing images with panoramic resolution of 2 m and multispectral resolution of 8 m.

12.1.6 Perform color compositions of CBERS-04A data

The CBERS-04A monitoring data from the WPM camera are used as an example of panchromatic band mapping at 2-m spatial resolution in monitoring performed on August 15, 2021 in the region of the Funil dam. The data obtained refer to processing level 4 (L4), referring to an orthorectified image with radiometric and geometric corrections using control points and a digital elevation model. More details on how to obtain data can be checked on a **website**[1] on the Internet.

In this case, the software R is used as an example; however, other programs can be used to accomplish this task at the student's discretion.

12.1.7 Enabling packages

The R packages `raster` (Hijmans et al., 2020), `RStoolbox` (Leutner et al., 2019), `xROI` (Seyednasrollah et al., 2021), `terra` (Hijmans et al., 2022), `rgdal` (Bivand et al., 2021), `uavRst` (Reudenbach et al., 2022), `lattice` (Sarkarand et al., 2021), `ggplot2` (Wickham et al., 2022), `gridExtra` (Auguie & Antonov, 2017), and `randomcoloR` (Ammar, 2019) are used for analysis and they are enabled for use with the `library` function.

```
library(raster)
library(RStoolbox)
library(xROI)
library(terra)
library(rgdal)
library(uavRst)
library(lattice)
library(ggplot2)
library(gridExtra)
library(randomcoloR)
```

[1] https://grupoqd.com.br/aquisicao-dos-dados-do-cbers-04a/

12.1.8 Import multispectral bands

The multispectral data for the CBERS-04A panchromatic (pan), blue, green, red, and near-infrared (NIR) bands are imported into R with the `raster` function from the directory in which the data are stored on the computer.

```
pan <- raster("C:/CBERS_4A_WPM_20210815_201_140_L4_BAND0.tif")
blue <- raster("C:/CBERS_4A_WPM_20210815_201_140_L4_BAND1.tif")
green <- raster("C:/CBERS_4A_WPM_20210815_201_140_L4_BAND2.tif")
red <- raster("C:/CBERS_4A_WPM_20210815_201_140_L4_BAND3.tif")
nir <- raster("C:/CBERS_4A_WPM_20210815_201_140_L4_BAND4.tif")
```

The multispectral bands in raster layers are stacked, cropped, and renamed to an extent of interest in the neighborhood of the Federal University of Lavras campus, state of Minas Gerais, Brazil, with the `stack`, `extent`, `crop`, and `names` functions, respectively.

```
c4multi<-stack(blue, green, red, nir) # stack
e<-extent(501789.4, 503332, 7652923, 7654290) # extent
c4m<-crop(c4multi, e) # crop
names(c4m)<-c("blue", "green", "red", "nir")
c4m
#class       : RasterBrick
#dimensions : 170, 193, 32810, 4  (nrow, ncol, ncell, nlayers)
#resolution : 8, 8  (x, y)
#extent      : 501788, 503332, 7652926, 7654286  (xmin, xmax, ymin, ymax)
#crs         : +proj=utm +zone=23 +south +datum=WGS84 +units=m +no_defs
#source      : memory
#names       : blue, green, red, nir
#min values :  156,    95,  87,  74
#max values :  641,   619, 730, 699
```

12.1.9 Performing natural color and false-color mapping

The `ggRGB` function is used to perform natural-color (R3G2B1) and infrared false-color (R4G3B2) mappings of the CBERS-04A imagery in the neighborhood of the Federal University of Lavras campus. Linear enhancement is applied in both situations with the `stretch="lin"` argument.

```
ggRGB(c4m, r = 3, g = 2, b = 1, stretch = "lin") + ggtitle("R3G2B1")
ggRGB(c4m, r = 4, g = 3, b = 2, stretch = "lin") + ggtitle("R4G3B2")
```

The use of the infrared false-color composition is extremely useful to eliminate doubts about the identity of vegetation and water targets monitored in this spatial subset of the CBERS-04A scene obtained in the Lavras region (Figure 12.1).

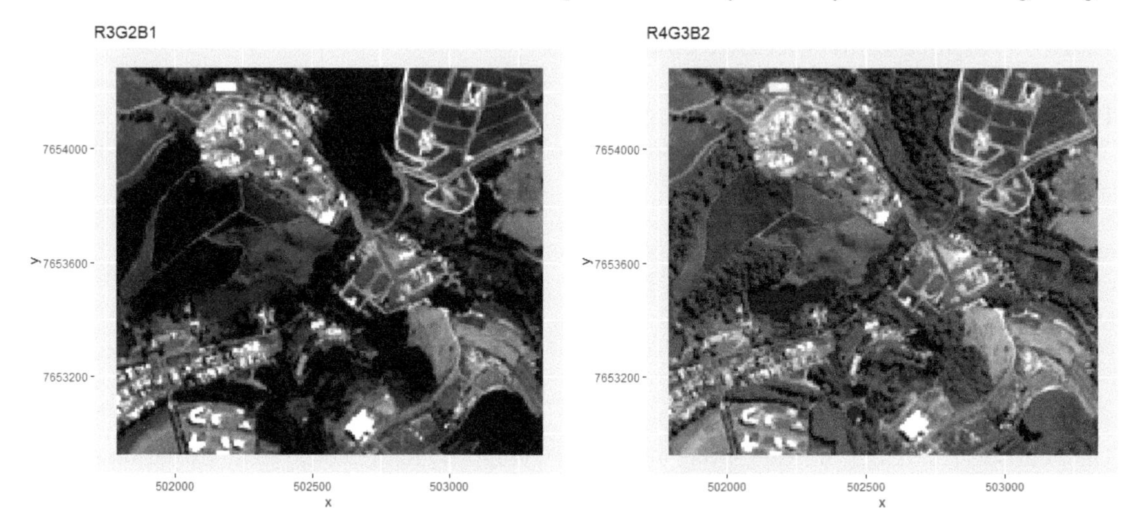

FIGURE 12.1 Mapping of natural color (left) and false-color (right) compositions from the CBERS-04A satellite imagery in the Lavras region, state of Minas Gerais, on August 15, 2021.

12.1.10 Perform fusion between color compositions and the panchromatic band

12.1.10.1 Crop the panchromatic band to the same extent as the multispectral band

The `crop` function is used to crop the panchromatic band to the same geographic region as the color compositions in the previous example. The `names` function is used to rename the raster layer as `pan`.

```
c4p<-crop(pan, c4m)
names(c4p)<-c("pan")
```

12.1.10.2 Fusion

The `panSharpen` function is used to perform the fusion between the panchromatic and multispectral bands of the CBERS-4A satellite in natural color (R3G2B1) and false-color (R4G3B2) compositions.

```
c4brovey <- panSharpen(c4m, c4p, r = 3, g = 2, b = 1, method = "brovey")
c4broveyI <- panSharpen(c4m, c4p, r = 4, g = 3, b = 2, method = "brovey")
```

The `ggRGB` function is used for mapping the analyzed region into natural color and false-color compositions of the multispectral bands under Brovey fusion.

```
ggRGB(c4brovey, r = 3, g = 2, b = 1, stretch = "lin") +
    ggtitle("fusão pan e R3G2B1")
ggRGB(c4broveyI, r = 4, g = 3, b = 2, stretch = "lin") +
    ggtitle("fusão pan e R4G3B2")
```

The targets under fusion are apparently sharper in the color compositions when compared to the non-fusion compositions. In the infrared color compositions the effect of the original color modification by the fusion process is minimized (Figure 12.2).

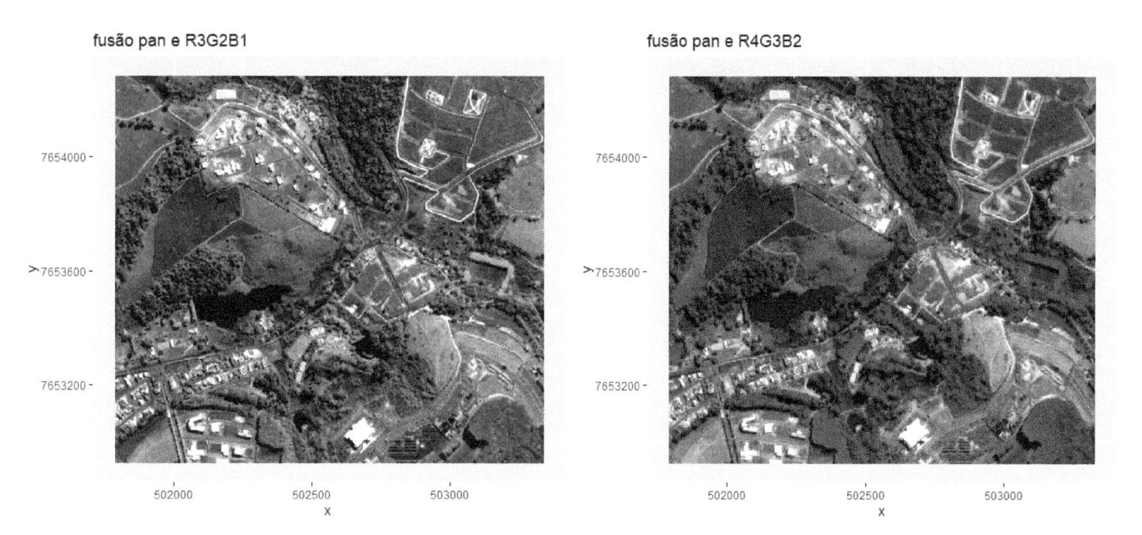

FIGURE 12.2 Mapping natural color (left) and false-color (right) compositions from CBERS-04A satellite imagery in the Lavras region, state of Minas Gerais, on August 15, 2021, after fusion of the multispectral and panchromatic bands by the Brovey method.

A single file with the fused bands is created by adding the fused NIR band to the dataset with the fused RGB bands using the `addLayer` function.

```
c4f <- addLayer(c4brovey, c(c4broveyI$nir_pan))
```

The non-fused multispectral bands are resampled to the same resolution as the panchromatic band by the nearest neighbor method to enable stacking the fused and non-fused images in a single file. The nearest neighbor interpolation method is used to have little modification of the original data by improving the spatial resolution of the raster from 8 to 2 m. Rasters are transformed into the `SpatRaster` class to perform this operation with the `rast` function.

```
c4fras<-rast(c4f) # Transform into SpatRaster class
c4mras<-rast(c4m) # Transform into SpatRaster class
c4mres <- terra::resample(c4mras, c4fras, method = "near") # Resample
```

The resampled data is converted back to the `raster` class to generate a single data file with both fused and non-fused images.

```
c4mresR <- as(c4mres, "Raster")
```

The file with all the data is obtained by adding the original multispectral, fused and panchromatic images into a single `RasterStack` class file.

```
c4mpf<-addLayer(c4mresR, c(c4f,c4p))
```

The result is exported into a known directory for later use with the `writeRaster` function.

```
writeRaster(c4mpf, filename="C:/sr/c11/c4mpf.tif",
            options="INTERLEAVE=BAND", overwrite=TRUE)
```

The file ready for exploratory data analysis is imported into R with the `stack` function. The `names` function is used to rename each stacked raster layer.

```
c4mpf <- stack("C:/sr/c11/c4mpf.tif")
names(c4mpf) <- c("blue","green","red","nir","blue_pan","green_pan","red_pan",
                  "nir_pan","pan")
```

Two files are produced, the first with the multispectral and panchromatic bands, and the second with the fused and panchromatic bands.

```
c4<- c4mpf[[c(1:4,9)]]
c4pan<-c4mpf[[c(5:9)]]
```

From the separate files, multispectral indices and RGB indices are determined from the imaging data. The imaging data are divided by 1000 and the `spectralIndices` and `rgb_indices` functions are applied to the data to obtain the indices from the multispectral bands. The indices are stacked into a single file with the `stack` function and exported to a directory on the computer with the `writeRaster` function.

```
# Scale factor
c4si<-c4/1000
# Spectral indices
SIc4 <- spectralIndices(c4si, blue = "blue", green= "green",
red = "red", nir = "nir",
scaleFactor = 1,
coefs = list(L = 0.5, G = 2.5, L_evi = 1, C1 = 6, C2 = 7.5, s = 1))
## RGB indices
c4rgb<-rgb_indices(red = c4si[[3]], green = c4si[[2]], blue  = c4si[[1]],
    rgbi = c("VVI", "VARI", "NDTI", "RI", "SCI", "BI", "SI", "HI", "TGI", "GLI",
             "NGRDI","GRVI", "GLAI", "HUE", "CI", "SAT", "SHP"))
# Stack data
c4SI<-stack(c4, SIc4, c4rgb)
# Export data
writeRaster(c4SI, filename="C:/sr/c11/c4SI.tif",
            options="INTERLEAVE=BAND", format="GTiff", overwrite=TRUE)
```

The same procedure as above is applied to the fused bands.

```
# Scale factor
c4sipan <- c4pan/1000
# Spectral indices
```

```
SIc4pan <- spectralIndices(c4sipan, blue = "blue_pan", green= "green_pan",
red = "red_pan", nir = "nir_pan",
scaleFactor = 1,
coefs = list(L = 0.5, G = 2.5, L_evi = 1, C1 = 6, C2 = 7.5, s = 1))
## RGB indices
c4rgbpan <- rgb_indices(red = c4sipan[[3]], green = c4sipan[[2]],
                         blue = c4sipan[[1]],
    rgbi = c("VVI", "VARI", "NDTI", "RI", "SCI", "BI", "SI", "HI", "TGI", "GLI",
    "NGRDI", "GRVI", "GLAI", "HUE", "CI", "SAT", "SHP"))
# Stack data
c4SIpan<-stack(c4pan, SIc4pan, c4rgbpan)
# Export data
writeRaster(c4SIpan, filename="C:/sr/c11/c4SIpan.tif",
            options="INTERLEAVE=BAND", format="GTiff", overwrite=TRUE)
```

The big data prepared in relation to the CBERS-04A multispectral imagery is imported into R with the `stack` function. The raster layers are renamed with the `names` function.

```
c4SI <- stack("C:/sr/c11/c4SI.tif")
names(c4SI) <- c("blue","green","red","nir","pan","CTVI","DVI","EVI","EVI2",
"GEMI","GNDVI","MSAVI","MSAVI2","NDVI","NDWI","NRVI","RVI","SAVI","SR","TTVI",
"TVI","WDVI","VVI","VARI","NDTI","RI","SCI","BI","SI","HI","TGI","GLI","NGRDI",
"GRVI","GLAI","HUE","CI","SAT","SHP")
```

The big data prepared from the fused multispectral imaging is imported into R with the `stack` function. The raster layers are renamed with the `names` function.

```
c4SIpan <- stack("C:/sr/c11/c4SIpan.tif")
names(c4SIpan) <- c("blue_pan","green_pan","red_pan","nir_pan","pan","CTVI",
"DVI","EVI","EVI2","GEMI","GNDVI","MSAVI","MSAVI2","NDVI","NDWI","NRVI","RVI",
"SAVI","SR","TTVI","TVI","WDVI","VVI","VARI","NDTI","RI","SCI","BI","SI","HI",
"TGI","GLI","NGRDI","GRVI","GLAI","HUE","CI","SAT","SHP")
```

Thereafter this data is used in the remaining steps of obtaining training samples, performing exploratory analysis of targets obtained from the training samples and imagery data, and supervised classification for recognition of targets in the region of interest.

12.1.11 Create spatial polygons with attributes

A natural color RGB composite is performed with the fused images to facilitate the creation of spatial polygons with attributes referring to each target analyzed using more detailed information.

```
plotRGB(c4SIpan, r=3, g=2, b=1, axes=F, stretch="lin")
```

In general, the visually recognized targets are: building (1), water (2), bare soil (3), mixed forest (4), giant bamboo (5), eucalyptus (6), mahogany (7), pasture (8), coffee crop (9), and solar panel (10). A `data.frame` class attribute database is created with three polygon repetitions for each of these classes observed in the image using the `xROI` package. The polygons are created interactively by being drawn above each target in the image. Data collection is terminated by choosing `finish` from the interactive menu or by typing `Esc` on the keyboard.

```
edif1 <- if(interactive()){drawPolygon()}
edif2 <- if(interactive()){drawPolygon()}
edif3 <- if(interactive()){drawPolygon()}
agua1 <- if(interactive()){drawPolygon()}
agua2 <- if(interactive()){drawPolygon()}
agua3 <- if(interactive()){drawPolygon()}
solo1 <- if(interactive()){drawPolygon()}
solo2 <- if(interactive()){drawPolygon()}
solo3 <- if(interactive()){drawPolygon()}
flor1 <- if(interactive()){drawPolygon()}
flor2 <- if(interactive()){drawPolygon()}
flor3 <- if(interactive()){drawPolygon()}
bambu1 <- if(interactive()){drawPolygon()}
bambu2 <- if(interactive()){drawPolygon()}
bambu3 <- if(interactive()){drawPolygon()}
euca1 <- if(interactive()){drawPolygon()}
euca2 <- if(interactive()){drawPolygon()}
euca3 <- if(interactive()){drawPolygon()}
mogno1 <- if(interactive()){drawPolygon()}
mogno2 <- if(interactive()){drawPolygon()}
mogno3 <- if(interactive()){drawPolygon()}
past1 <- if(interactive()){drawPolygon()}
past2 <- if(interactive()){drawPolygon()}
past3 <- if(interactive()){drawPolygon()}
cafe1 <- if(interactive()){drawPolygon()}
cafe2 <- if(interactive()){drawPolygon()}
cafe3 <- if(interactive()){drawPolygon()}
placa1 <- if(interactive()){drawPolygon()}
placa2 <- if(interactive()){drawPolygon()}
placa3 <- if(interactive()){drawPolygon()}
```

The spatial polygons are created in a single `p` file with identifying codes for each sample collected in the image with the `SpatialPolygons` function, totaling 30 polygons and 10 distinct classes.

```
(p <- SpatialPolygons(list(Polygons(list(Polygon(cbind(edif1))), "1"),
     Polygons(list(Polygon(cbind(edif2))), "2"),
     Polygons(list(Polygon(cbind(edif3))), "3"),
     Polygons(list(Polygon(cbind(agua1))), "4"),
     Polygons(list(Polygon(cbind(agua2))), "5"),
     Polygons(list(Polygon(cbind(agua3))), "6"),
     Polygons(list(Polygon(cbind(solo1))), "7"),
     Polygons(list(Polygon(cbind(solo2))), "8"),
     Polygons(list(Polygon(cbind(solo3))), "9"),
     Polygons(list(Polygon(cbind(flor1))), "10"),
     Polygons(list(Polygon(cbind(flor2))), "11"),
```

```
    Polygons(list(Polygon(cbind(flor3))), "12"),
    Polygons(list(Polygon(cbind(bambu1))), "13"),
    Polygons(list(Polygon(cbind(bambu2))), "14"),
    Polygons(list(Polygon(cbind(bambu3))), "15"),
    Polygons(list(Polygon(cbind(euca1))), "16"),
    Polygons(list(Polygon(cbind(euca2))), "17"),
    Polygons(list(Polygon(cbind(euca3))), "18"),
    Polygons(list(Polygon(cbind(mogno1))), "19"),
    Polygons(list(Polygon(cbind(mogno2))), "20"),
    Polygons(list(Polygon(cbind(mogno3))), "21"),
    Polygons(list(Polygon(cbind(past1))), "22"),
    Polygons(list(Polygon(cbind(past2))), "23"),
    Polygons(list(Polygon(cbind(past3))), "24"),
    Polygons(list(Polygon(cbind(cafe1))), "25"),
    Polygons(list(Polygon(cbind(cafe2))), "26"),
    Polygons(list(Polygon(cbind(cafe3))), "27"),
    Polygons(list(Polygon(cbind(placa1))), "28"),
    Polygons(list(Polygon(cbind(placa2))), "29"),
    Polygons(list(Polygon(cbind(placa3))), "30"))) )
```

A data frame with line and class identifier code for each polygon is created according to the total length of polygons.

```
(p.df <- data.frame(ID=1:length(p)))
#   ID
#1   1
#2   2
#3   3
#4   4
#5   5
#6   6
#7   7
#8   8
#9   9
#10 10
#11 11
#12 12
#13 13
#14 14
#15 15
#16 16
#17 17
#18 18
#19 19
#20 20
#21 21
#22 22
#23 23
#24 24
#25 25
#26 26
```

```
#27 27
#28 28
#29 29
#30 30
rownames(p.df)
#[1] "1"  "2"  "3"  "4"  "5"  "6"  "7"  "8"  "9"  "10" "11" "12" "13" "14" "15"
"16" "17" "18" "19" "20" "21" "22" "23"
#[24] "24" "25" "26" "27" "28" "29" "30"
```

The previously created database is associated with the polygons to generate spatial polygons with attributes using the `SpatialPolygonsDataFrame` function.

```
pol <- SpatialPolygonsDataFrame(p, p.df)
```

The column with the identifier code for each class is modified in the polygon attribute database according to the class of each polygon. Since the class of the first polygon already has the code 1, it is not necessary to modify the class of the first polygon, and the method starts from the second polygon created.

```
pol[pol$ID == 2,"class"] <- 1
pol[pol$ID == 3,"class"] <- 1
pol[pol$ID == 4,"class"] <- 2
pol[pol$ID == 5,"class"] <- 2
pol[pol$ID == 6,"class"] <- 2
pol[pol$ID == 7,"class"] <- 3
pol[pol$ID == 8,"class"] <- 3
pol[pol$ID == 9,"class"] <- 3
pol[pol$ID == 10,"class"] <- 4
pol[pol$ID == 11,"class"] <- 4
pol[pol$ID == 12,"class"] <- 4
pol[pol$ID == 13,"class"] <- 5
pol[pol$ID == 14,"class"] <- 5
pol[pol$ID == 15,"class"] <- 5
pol[pol$ID == 16,"class"] <- 6
pol[pol$ID == 17,"class"] <- 6
pol[pol$ID == 18,"class"] <- 6
pol[pol$ID == 19,"class"] <- 7
pol[pol$ID == 20,"class"] <- 7
pol[pol$ID == 21,"class"] <- 7
pol[pol$ID == 22,"class"] <- 8
pol[pol$ID == 23,"class"] <- 8
pol[pol$ID == 24,"class"] <- 8
pol[pol$ID == 25,"class"] <- 9
pol[pol$ID == 26,"class"] <- 9
pol[pol$ID == 27,"class"] <- 9
pol[pol$ID == 28,"class"] <- 10
pol[pol$ID == 29,"class"] <- 10
pol[pol$ID == 30,"class"] <- 10
```

For the purpose of comparing results in a more generalist approach, a simplified version of classes is performed by grouping the previous database into the classes: building (1), water (2), exposed soil (3) and vegetation (4). In this case, the class `solar plate` is grouped into `building` and `mixed forest`, `giant bamboo`, `eucalyptus`, `mahogany`, `pasture`, and `coffee`, into the class `vegetation`.

```
pol1<-pol
pol1[pol1$ID == 2,"class"] <- 1
pol1[pol1$ID == 3,"class"] <- 1
pol1[pol1$ID == 4,"class"] <- 2
pol1[pol1$ID == 5,"class"] <- 2
pol1[pol1$ID == 6,"class"] <- 2
pol1[pol1$ID == 7,"class"] <- 3
pol1[pol1$ID == 8,"class"] <- 3
pol1[pol1$ID == 9,"class"] <- 3
pol1[pol1$ID == 10,"class"] <- 4
pol1[pol1$ID == 11,"class"] <- 4
pol1[pol1$ID == 12,"class"] <- 4
pol1[pol1$ID == 13,"class"] <- 4
pol1[pol1$ID == 14,"class"] <- 4
pol1[pol1$ID == 15,"class"] <- 4
pol1[pol1$ID == 16,"class"] <- 4
pol1[pol1$ID == 17,"class"] <- 4
pol1[pol1$ID == 18,"class"] <- 4
pol1[pol1$ID == 19,"class"] <- 4
pol1[pol1$ID == 20,"class"] <- 4
pol1[pol1$ID == 21,"class"] <- 4
pol1[pol1$ID == 22,"class"] <- 4
pol1[pol1$ID == 23,"class"] <- 4
pol1[pol1$ID == 24,"class"] <- 4
pol1[pol1$ID == 25,"class"] <- 4
pol1[pol1$ID == 26,"class"] <- 4
pol1[pol1$ID == 27,"class"] <- 4
pol1[pol1$ID == 28,"class"] <- 1
pol1[pol1$ID == 29,"class"] <- 1
pol1[pol1$ID == 30,"class"] <- 1
```

The same coordinate system as in the image, Universal Transverse Mercator (UTM), zone 23 South, is assigned to the polygons with multiple classes (pol) and with generalized classes (pol1) configuring the `+proj` argument.

```
crs(pol) <- "+proj=utm +zone=23 +south +datum=WGS84 +units=m +no_defs"
crs(pol1) <- "+proj=utm +zone=23 +south +datum=WGS84 +units=m +no_defs"
```

The spatial polygons with created attributes are mapped comparatively over the natural color composition of the CBERS-04A imagery.

```
par(mfrow=c(1,2), mar = c(1, 1, 0.3, 0.3), mgp = c(1.5, 0.6, 0))
colors <- c("white", "blue", "brown", "green", "SpringGreen", "PaleGreen",
            "DarkOliveGreen", "YellowGreen", "DarkGreen", "black")
plotRGB(c4SIpan, r=3, g=2, b=1, axes=F, stretch="lin", margins=T)
plot(pol, add=T, col=colors[pol$class], pch = 19)
color <- c("white", "blue", "brown", "green")
```

```
plotRGB(c4SIpan, r=3, g=2, b=1, axes=F, stretch="lin", margins=T)
plot(pol1, add=T, col=color[pol1$class], pch = 19)
```

Thus, it is possible to check the generalizations of more detailed (left) and less detailed (right) classes (Figure 12.3).

FIGURE 12.3 Natural color composition mapping with detailed polygons (left) and with generalized polygons (right) from the CBERS-04A imagery in the Lavras region, state of Minas Gerais, on August 15, 2021, after fusion of the multispectral bands with the panchromatic band by the Brovey method.

The polygons created are exported for later use with the `writeOGR` function.

```
writeOGR(pol,dsn="C:/geo/c11/pol.shp","pol",driver="ESRI Shapefile") # pol
writeOGR(pol1,dsn="C:/geo/c11/pol1.shp","pol1",driver="ESRI Shapefile") # pol1
```

The `readOGR` function is used to import the spatial polygons with created attributes.

12.1.12 Perform exploratory analysis

To perform the exploratory analysis, the polygons are transformed into raster data using the `rasterize` function. This is performed for all 4 evaluated treatments of multispectral images with and without class detail and fused images with and without class detail.

```
pol_raster <- rasterize(pol, c4SI, field = "class")
pol1_raster <- rasterize(pol1, c4SI, field = "class")
pol_raster_pan <- rasterize(pol, c4SIpan, field = "class")
pol1_raster_pan <- rasterize(pol1, c4SIpan, field = "class")
```

Raster polygons are mapped with the `plot` function.

The `zonal` function with the `fun="mean"` argument is used to determine the mean pixel value of the CBERS-04A imagery with and without fusion for each polygon class, considering the approaches with 10 and 4 polygon classes.

```
pol_zonal <- zonal(c4SI, pol_raster, fun="mean", na.rm=TRUE)
pol1_zonal <- zonal(c4SI, pol1_raster, fun="mean", na.rm=TRUE)
pol_zonal_pan <- zonal(c4SIpan, pol_raster_pan, fun="mean", na.rm=TRUE)
pol1_zonal_pan <- zonal(c4SIpan, pol1_raster_pan, fun="mean", na.rm=TRUE)
```

The table with the zonal statistics is transposed to plot the variables against the targets sampled by polygons with different class generalizations for each type of imaging data considered in the analysis.

```
signature <- t(pol_zonal)
signature1 <- t(pol1_zonal)
signaturep <- t(pol_zonal_pan)
signature1p <- t(pol1_zonal_pan)
```

Exploratory barplot analysis is used to evaluate the magnitude of values sampled by polygons in the approaches with 10 and 4 classes of target types analyzed considering the range of values of all bands.

The `barplot` function is used to plot the magnitude of each spectral band evaluated against polygons with 10 targets and 4 targets under multispectral imaging and fusion.

```
par(mfrow = c(4, 1),mar = c(3.5, 4.0, 1, 1), mgp = c(2.0, 0.5, 0.0))
# Multispectral (10 classes)
colors <- c("blue","green","red","brown")
Target <- c('1','2','3','4','5','6','7','8','9','10')
bands <- c("blue","green","red","nir")
Values <- matrix(c(signature[2:5,]), nrow = 4, ncol = 10, byrow = F)
barplot(Values, main = "Multispectral (10 classes)", names.arg = Target,
        xlab = "Target", ylab = "Spectral target information", col = colors,
        ylim=c(0,2000))
legend("topleft", bands, cex = 0.9, fill = colors, xpd=TRUE, inset=c(0.7,0.0),
       horiz=TRUE)
# Multispectral (4 classes)
colors <- c("blue","green","red","brown")
Target <- c('1','2','3','4')
bands <- c("blue","green","red","nir")
Values <- matrix(c(signature1[2:5,]), nrow = 4, ncol = 4, byrow = F)
barplot(Values, main = "Multispectral (4 classes)", names.arg = Target,
        xlab = "Target", ylab = "Spectral target information", col = colors,
        ylim=c(0,1500))
# Fusion (10 classes)
colors <- c("blue","green","red","brown")
Target <- c('1','2','3','4','5','6','7','8','9','10')
bands <- c("blue","green","red","nir")
Values <- matrix(c(signaturep[2:5,]), nrow = 4, ncol = 10, byrow = F)
barplot(Values, main = "Fusion (10 classes)", names.arg = Target,
        xlab = "Target", ylab = "Spectral target information", col = colors,
```

```
        ylim=c(0,900))
# Fusion (4 classes)
colors <- c("blue","green","red","brown")
Target <- c('1','2','3','4')
bands <- c("blue","green","red","nir")
Values <- matrix(c(signature1p[2:5,]), nrow = 4, ncol = 4, byrow = F)
barplot(Values, main = "Fusion (4 classes)", names.arg = Target,
        xlab = "Target", ylab = "Spectral target information", col = colors,
        ylim=c(0,500))
```

With the barplot it is possible to identify that there is magnitude variation according to the landscape target analyzed, as well as the image fusion processing (Figure 12.4).

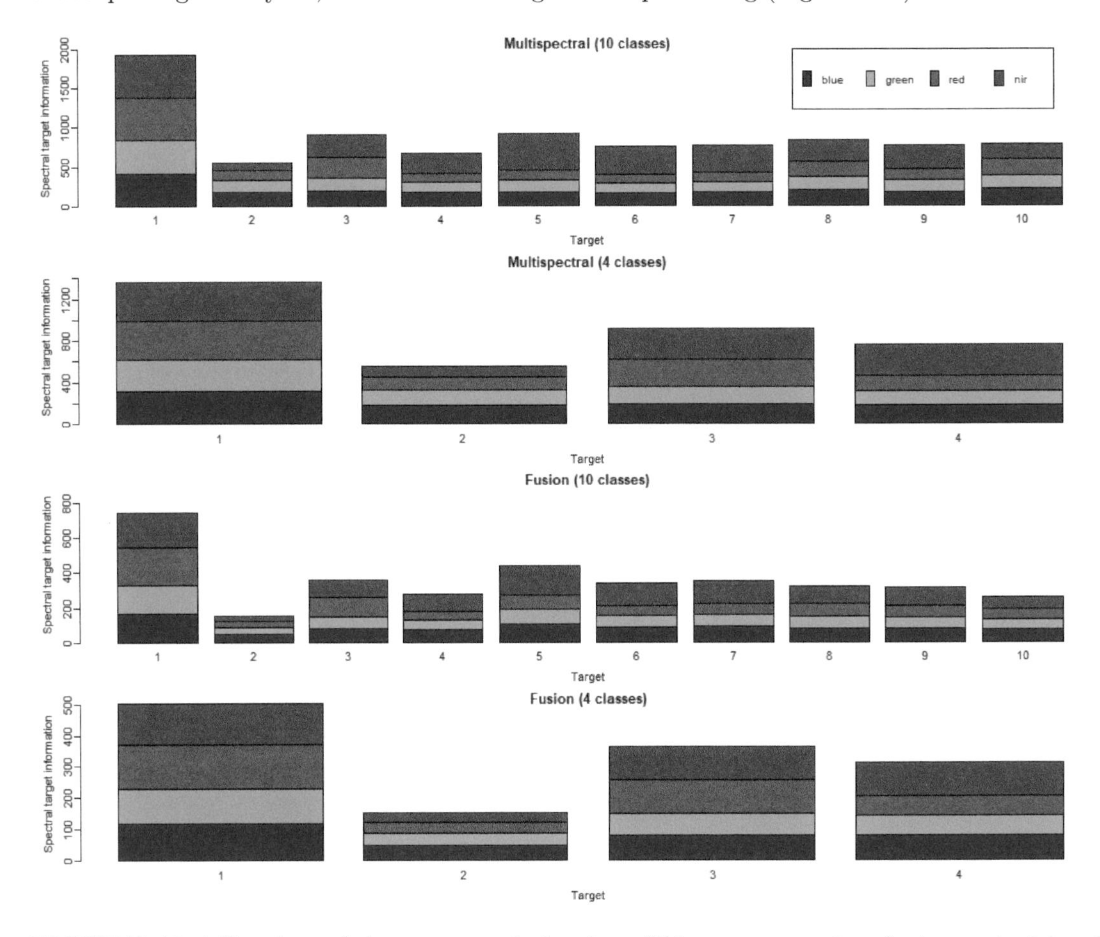

FIGURE 12.4 Barplots of the average pixel value of blue, green, red and nir spectral bands sampled by spatial polygons with attributes in a classification experiment considering 10 classes and 4 classes, and image processing with and without spectral band fusion, in the region of Lavras, state of Minas Gerais, on August 15, 2021.

The determination of exploratory analysis of spectral indices can be more complex in view of the larger number of indices determined in the analysis. The `randomcolor` package is used to obtain a random color palette in the hexadecimal system to graphically represent each of the 33 analyzed indices.

```
n <- 33
palette <- distinctColorPalette(n)
palette
#[1] "#A52CF1" "#E85578" "#868B73" "#74B9BA" "#89EA88" "#5FA8E6" "#B1C2E0"
"#C87FE6" "#D8E48B" "#76E746" "#8153DE"
#[12] "#DF4DDE" "#DAE3B7" "#86AD51" "#E29DDA" "#60EDB3" "#5BBD95" "#E0BDD6"
"#D2E8EB" "#E3733C" "#E2CDC2" "#D9A97B"
#[23] "#D2E74C" "#74EAE6" "#74789B" "#7372D0" "#DD8C8F" "#A9A4E4" "#9E5790"
"#63CAE6" "#E75FB5" "#A4EFCC" "#E6BB51"
```

The `barplot` function is used to plot the magnitude of each index obtained from spectral bands evaluated against polygons with 10 targets and 4 targets under multispectral imaging and fusion. A series of functions are developed below to compare the bar plots relative to each of the treatments performed.

```
par(mfrow = c(4, 1),mar = c(3.5, 4.0, 1, 5), mgp = c(2.0, 0.5, 0.0))
colors <- palette
Target <- c('1','2','3','4','5','6','7','8','9','10')
Target4 <- c('1','2','3','4')
bands <- c("CTVI","DVI","EVI","EVI2","GEMI","GNDVI","MSAVI","MSAVI2","NDVI",
  "NDWI","NRVI","RVI","SAVI","SR","TTVI","TVI","WDVI","VVI","VARI","NDTI","RI",
  "SCI","BI","SI","HI","TGI","GLI","NGRDI","GRVI","GLAI","HUE","CI","SAT","SHP")
# Multispectral (10 classes)
Values <- matrix(c(signature[7:39,]), nrow = 33, ncol = 10, byrow = F)
barplot(Values, main = "Multispectral (10 classes)", names.arg = Target,
        xlab = "Target", ylab = "Spectral target information", col = colors,
        ylim=c(0,100))
legend("topright", bands, cex = 0.8, fill = colors, ncol=1, xpd=NA,
       inset=c(-0.06,0.0))
# Multispectral (4 classes)
Values <- matrix(c(signature1[7:39,]), nrow = 33, ncol = 4, byrow = F)
barplot(Values, main = "Multispectral (4 classes)", names.arg = Target4,
        xlab = "Target", ylab = "Spectral target information", col = colors,
        ylim=c(0,100))
# Fusion (10 classes)
Values <- matrix(c(signaturep[7:39,]), nrow = 33, ncol = 10, byrow = F)
barplot(Values, main = "Fusion (10 classes)", names.arg = Target,
        xlab = "Target", ylab = "Spectral target information", col = colors,
        ylim=c(0,550))
# Fusion (4 classes)
Values <- matrix(c(signature1p[7:39,]), nrow = 33, ncol = 4, byrow = F)
barplot(Values, main = "Fusion (4 classes)", names.arg = Target4,
        xlab = "Target", ylab = "Spectral target information", col = colors,
        ylim=c(0,500))
```

With the barplot it is possible to identify that there is variation of the index magnitude according to the analyzed landscape target, as well as the image fusion processing. The indices with the greatest magnitude of variation are RI, HI and GLAI (Figure 12.5).

The spectral signature of the targets is performed continuously to compare the spectral response of each of the treatments evaluated in the experiment. The codes used are presented in sequential form to prepare the graphs for the treatments performed.

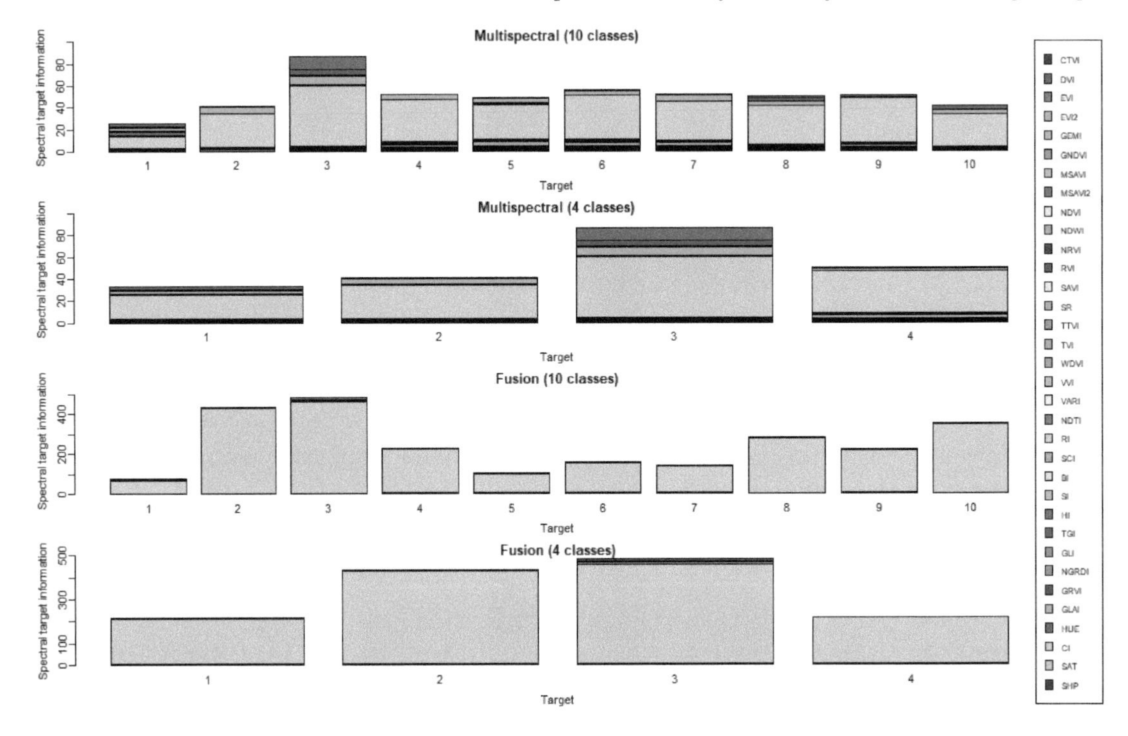

FIGURE 12.5 Barplots of the mean pixel value of indices obtained from spectral bands sampled by spatial polygons with attributes in a classification experiment considering 10 classes and 4 classes, and image processing with and without spectral band fusion, in the region of Lavras, state of Minas Gerais, on August 15, 2021.

```
par(mfrow = c(2, 2),mar = c(3.5, 4.0, 2, 1), mgp = c(2.0, 0.5, 0.0))
# Multispectral (10 classes)
plot(0, ylim=c(0,1000), xlim = c(2.1,4.9), type='n', xlab="Bands",
     ylab = "Surface reflectance", main = "Multispectral (10 classes)")
lines(signature[,1], col="grey", lwd=3)
lines(signature[,2], col="blue", lwd=3)
lines(signature[,3], col="brown", lwd=3)
lines(signature[,4], col="ForestGreen", lwd=3)
lines(signature[,5], col="green", lwd=3)
lines(signature[,6], col="LightGreen", lwd=3)
lines(signature[,7], col="DarkOliveGreen", lwd=3)
lines(signature[,8], col="YellowGreen", lwd=3)
lines(signature[,9], col="SpringGreen", lwd=3)
lines(signature[,10], col="black", lwd=3)
legend("topleft", legend = c('1','2','3','4','5','6','7','8','9','10'), horiz=F,
  col=c("grey","blue","brown","ForestGreen","green","LightGreen",
 "DarkOliveGreen","YellowGreen","SpringGreen","black"),lty=1,lwd=3,
  cex = 0.8, ncol=5)
# Multispectral (4 classes)
plot(0, ylim=c(0,600), xlim = c(2.1,4.9), type='n', xlab="Bands",
     ylab = "Surface reflectance", main = "Multispectral (4 classes)")
lines(signature1[,1], col="grey", lwd=3)
lines(signature1[,2], col="blue", lwd=3)
lines(signature1[,3], col="brown", lwd=3)
```

```
lines(signature1[,4], col="ForestGreen", lwd=3)
legend("topleft", legend = c('1','2','3','4'), horiz=T,
       col=c("grey", "blue", "brown", "ForestGreen"), lty=1, lwd=3, cex = 0.8)
# Fusion (10 classes)
plot(0, ylim=c(0,350), xlim = c(2.1,4.9), type='n', xlab="Bands",
     ylab = "Surface reflectance", main = "Fusion (10 classes)")
lines(signaturep[,1], col="grey", lwd=3)
lines(signaturep[,2], col="blue", lwd=3)
lines(signaturep[,3], col="brown", lwd=3)
lines(signaturep[,4], col="ForestGreen", lwd=3)
lines(signaturep[,5], col="green", lwd=3)
lines(signaturep[,6], col="LightGreen", lwd=3)
lines(signaturep[,7], col="DarkOliveGreen", lwd=3)
lines(signaturep[,8], col="YellowGreen", lwd=3)
lines(signaturep[,9], col="SpringGreen", lwd=3)
lines(signaturep[,10], col="black", lwd=3)
# Fusion (4 classes)
plot(0, ylim=c(0,200), xlim = c(2.1,4.9), type='n', xlab="Bands",
     ylab = "Surface reflectance", main = "Fusion (4 classes)")
lines(signature1p[,1], col="grey", lwd=3)
lines(signature1p[,2], col="blue", lwd=3)
lines(signature1p[,3], col="brown", lwd=3)
lines(signature1p[,4], col="ForestGreen", lwd=3)
```

Based on the spectral signature plots, it is possible to observe that in general, the targets are well characterized in the region according to the training samples obtained, so that water has reflected the least energy since it is a body that absorbs a lot of energy when compared to other targets. On the other hand, urbanized areas reflect the most energy in all spectral bands. The exposed soil reflected more energy in the red (4) and near-infrared (5). Vegetation absorbed more energy in the visible (1 to 3) and reflected more energy in the near-infrared, and there was greater differentiation of distinct vegetation targets in the near-infrared (Figure 12.6).

12.1.13 Machine learning

Supervised classifications are performed with the random forest (`rf`), classification and regression trees (`CART`), `rpart1SE` and support vector machine `svmLinear2` algorithms.

The supervised classification is set up with the training data with 10 and 4 classes, respectively. The data was not partitioned in view of the small amount of sample repetition obtained for analysis. The method `kfold = 10` is used in the validation of the results. The function `superClass` is used to perform supervised classification.

```
# Multispectral (10 classes)
SC1a <- superClass(c4SI, trainData = pol, responseCol = "class",
model = "rf", kfold = 10)
SC2a <- superClass(c4SI, trainData = pol, responseCol = "class",
model = "rpart1SE", kfold = 10)
SC3a <- superClass(c4SI, trainData = pol, responseCol = "class",
model = "svmLinear2", kfold = 10)
# Multispectral (4 classes)
```

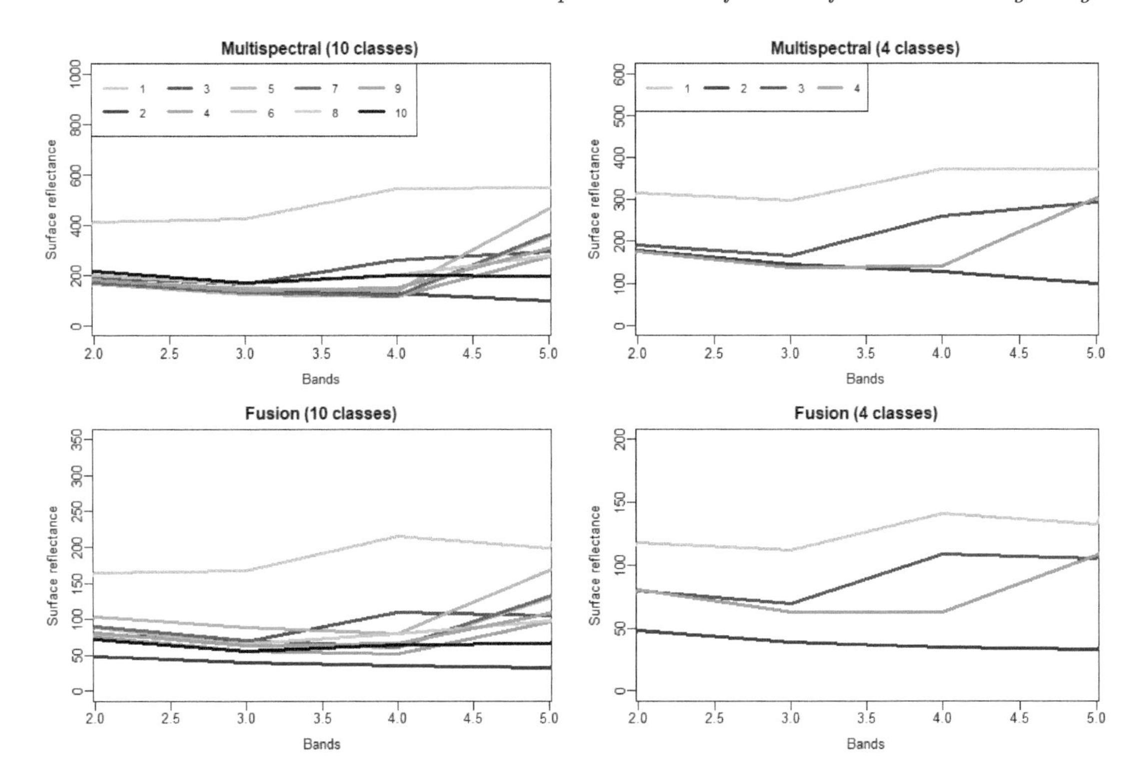

FIGURE 12.6 Spectral signature of bands sampled by spatial polygons with attributes in a classification experiment considering 10 and 4 classes, and image processing with and without spectral band fusion, in the region of Lavras, state of Minas Gerais, on August 15, 2021.

```
SC1b <- superClass(c4SI, trainData = pol1, responseCol = "class",
model = "rf", kfold = 10)
SC2b <- superClass(c4SI, trainData = pol1, responseCol = "class",
model = "rpart1SE", kfold = 10)
SC3b <- superClass(c4SI, trainData = pol1, responseCol = "class",
model = "svmLinear2", kfold = 10)
# Fusion (10 classes)
SC1c <- superClass(c4SIpan, trainData = pol, responseCol = "class",
model = "rf", kfold = 10)
SC2c <- superClass(c4SIpan, trainData = pol, responseCol = "class",
model = "rpart1SE", kfold = 10)
SC3c <- superClass(c4SIpan, trainData = pol, responseCol = "class",
model = "svmLinear2", kfold = 10)
# Fusion (4 classes)
SC1d <- superClass(c4SIpan, trainData = pol1, responseCol = "class",
model = "rf", kfold = 10)
SC2d <- superClass(c4SIpan, trainData = pol1, responseCol = "class",
model = "rpart1SE", kfold = 10)
SC3d <- superClass(c4SIpan, trainData = pol1, responseCol = "class",
model = "svmLinear2", kfold = 10)
```

The results obtained are exported to a directory on the computer for later use. The `saveRSTBX` function is used to export the classification results for each evaluated machine learning algorithm.

```
saveRSTBX(SC1a, filename="C:/sr/c10/new/SC1a", overwrite=TRUE)
saveRSTBX(SC2a, filename="C:/sr/c10/new/SC2a", overwrite=TRUE)
saveRSTBX(SC3a, filename="C:/sr/c10/new/SC3a", overwrite=TRUE)
saveRSTBX(SC1b, filename="C:/sr/c10/new/SC1b", overwrite=TRUE)
saveRSTBX(SC2b, filename="C:/sr/c10/new/SC2b", overwrite=TRUE)
saveRSTBX(SC3b, filename="C:/sr/c10/new/SC3b", overwrite=TRUE)
saveRSTBX(SC1c, filename="C:/sr/c10/new/SC1c", overwrite=TRUE)
saveRSTBX(SC2c, filename="C:/sr/c10/new/SC2c", overwrite=TRUE)
saveRSTBX(SC3c, filename="C:/sr/c10/new/SC3c", overwrite=TRUE)
saveRSTBX(SC1d, filename="C:/sr/c10/new/SC1d", overwrite=TRUE)
saveRSTBX(SC2d, filename="C:/sr/c10/new/SC2d", overwrite=TRUE)
saveRSTBX(SC3d, filename="C:/sr/c10/new/SC3d", overwrite=TRUE)
```

For the `rpart1SE` algorithm, a decision tree is used to represent the model's reasoning in predicting the classes as a function of the input spectral variables. The `plot` and `text` functions are used to determine the decision tree fitted by the `rpart1SE` model for each treatment.

```
par(mfrow = c(2, 2),mar = c(0, 0, 1.5, 0), mgp = c(2.0, 0.5, 0.0))
plot(SC2a$model$finalModel, uniform=TRUE, main = "Multispectral (10 classes)",
     margin=0.2)
text(SC2a$model$finalModel, digits = 3)
plot(SC2b$model$finalModel, uniform=TRUE, main = "Multispectral (4 classes)",
     ,margin=0.2)
text(SC2b$model$finalModel, digits = 3)
plot(SC2c$model$finalModel, uniform=TRUE, main = "Fusion (10 classes)",
     ,margin=0.2)
text(SC2c$model$finalModel, digits = 3)
plot(SC2d$model$finalModel, uniform=TRUE, main = "Fusion (10 classes)",
     ,margin=0.2)
text(SC2d$model$finalModel, digits = 3)
```

Note that the initial nodes of the 10-class classification with and without fusion are very similar, but after the second node differences begin to occur. The architecture in the multispectral approach is simpler than in the fusion case. In the case of the generalized 4-class classification, all decision nodes are different, but the network architecture is identical for both the multispectral and fusion approaches (Figure 12.7).

The Gini index is used to evaluate the importance of the variables in the random forest classification. The results of the Gini index are plotted with functions from the `ggplot2` package.

```
a <- ggplot(varImp(object=SC1a$model),
     top = dim(varImp(object=SC1a$model)$importance)[1],) +
     ggtitle("Multispectral (10 classes)")
b <- ggplot(varImp(object=SC1b$model),
     top = dim(varImp(object=SC1b$model)$importance)[1],) +
     ggtitle("Multispectral (4 classes)")
c <- ggplot(varImp(object=SC1c$model),
     top = dim(varImp(object=SC1c$model)$importance)[1],) +
     ggtitle("Fusion (10 classes)")
d <- ggplot(varImp(object=SC1d$model),
     top = dim(varImp(object=SC1d$model)$importance)[1],) +
```

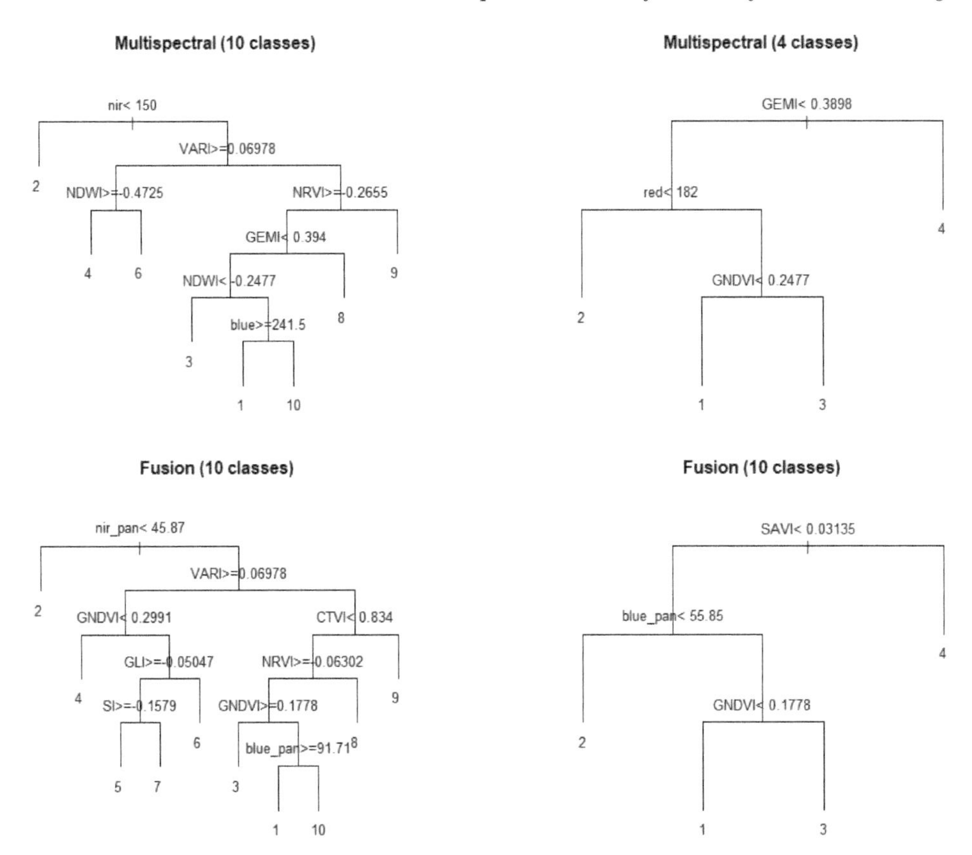

FIGURE 12.7 Decision trees defined by the machine learning algorithm `rpart1SE` after supervised classification considering 10 classes and 4 classes, and image processing with and without spectral band fusion, in the region of Lavras, state of Minas Gerais, on August 15, 2021.

```
    ggtitle("Fusion (4 classes)")
grid.arrange(a,b,c,d,ncol=2)
```

The importance of each big data variable used as input data in the classification varies with each treatment. However, in the case of fusion, the fused NIR band was the most important for both 10 and 4 training classes. In the case of non-fusion imaging data, the NDWI and GEMI indices were the most important for 10 and 4 classes, respectively (Figure 12.8).

Comparative mapping of the targets building (1), water (2), bare soil (3), mixed forest (4), giant bamboo (5), eucalyptus (6), mahogany (7), grassland (8), coffee crops (9) and solar plate (10) is performed with the functions `ggR`, `scale_fill_manual` and `grid.arrange`. In this first mapping the 10-class approach is analyzed.

```
cl<-c("grey92", "blue", "brown", "ForestGreen", "green", "LightGreen",
      "DarkOliveGreen", "YellowGreen", "SpringGreen", "black")
m1<-ggR(SC1a$map, geom_raster = TRUE)+
    scale_fill_manual(values = cl, breaks = 1:10,
    labels = c('1','2','3','4','5','6','7','8','9','10'),
    name = "A")
```

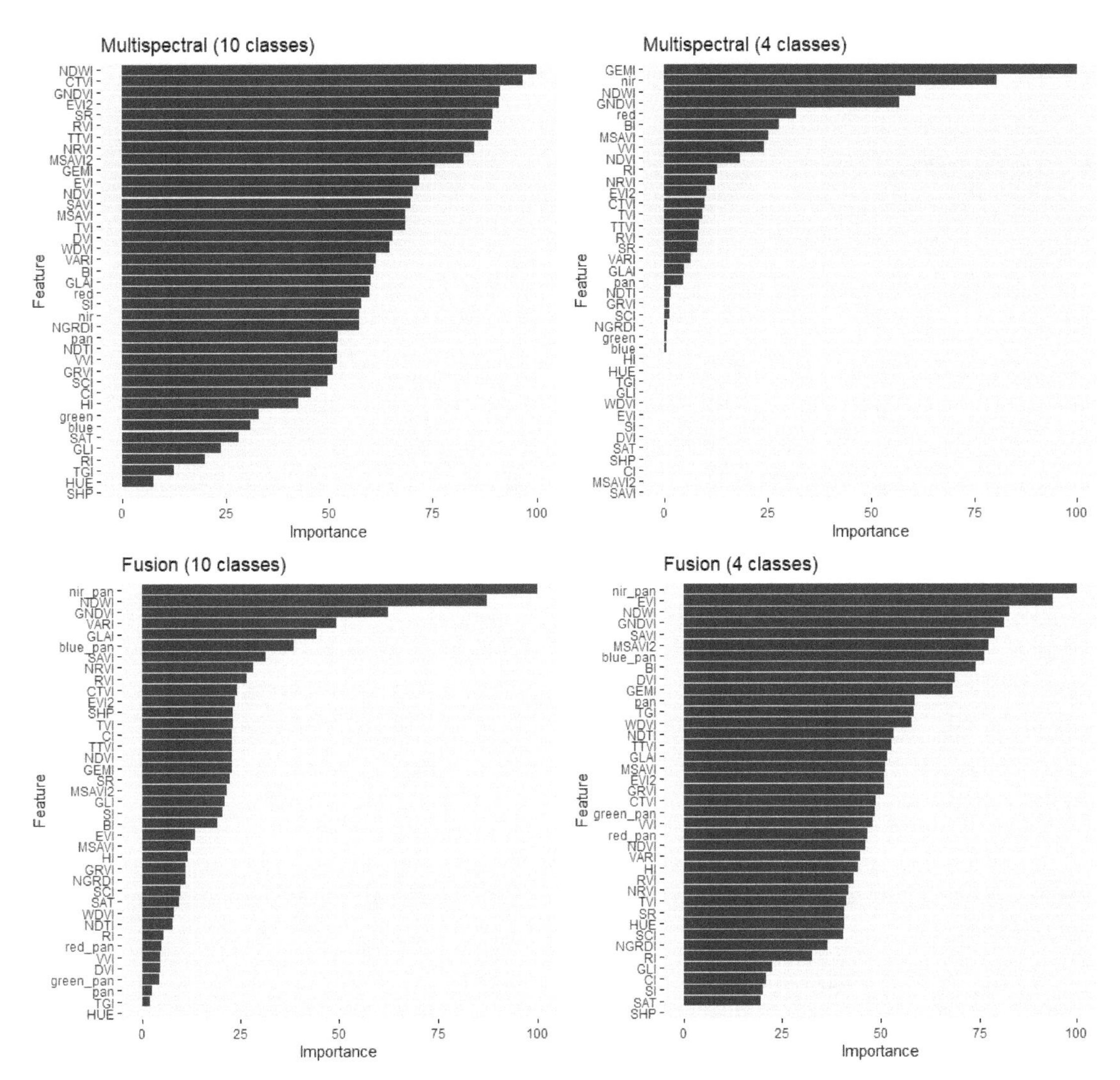

FIGURE 12.8 Importance of variables defined after training in supervised random forest classification considering 10 classes and 4 classes, and image processing with and without spectral band fusion, in the region of Lavras, state of Minas Gerais, on August 15, 2021.

```
m2<- ggR(SC2a$map, geom_raster = TRUE) +
    scale_fill_manual(values = cl, breaks = 1:10,
    labels = c('1','2','3','4','5','6','7','8','9','10'),
    name = "B")
m3<- ggR(SC3a$map, geom_raster = TRUE) +
    scale_fill_manual(values = cl, breaks = 1:10,
    labels = c('1','2','3','4','5','6','7','8','9','10'),
    name = "C")
m4<-ggR(SC1c$map, geom_raster = TRUE)+
    scale_fill_manual(values = cl, breaks = 1:10,
    labels = c('1','2','3','4','5','6','7','8','9','10'),
    name = "D")
m5<- ggR(SC2c$map, geom_raster = TRUE) +
```

```
    scale_fill_manual(values = cl, breaks = 1:10,
    labels = c('1','2','3','4','5','6','7','8','9','10'),
    name = "E")
m6<- ggR(SC3c$map, geom_raster = TRUE) +
    scale_fill_manual(values = cl, breaks = 1:10,
    labels = c('1','2','3','4','5','6','7','8','9','10'),
    name = "F")
grid.arrange(m1, m2, m3, m4, m5, m6, ncol=2)
```

The mapping results are visually similar. In both situations there were mapping errors of a solar panel that was mistaken for asphalt. There was also confusion of coffee crops with some types of grasslands and forest borders. The difference between multispectral data with and without fusion is complex to assess visually; however, in the case of the `svmLinear2` classification, fusion improved the interpretation of targets classified as solar plate when compared to `svmLinear2` without fusion. In the case of the `rf` and `rpart1SE` algorithms, the fusion process appeared to determine greater confusion of mapping solar plate instead of asphalt (Figure 12.9).

Comparative mapping of the targets building (1), water (2), bare soil (3), and vegetation (4) is performed with the functions `ggR`, `scale_fill_manual` and `grid.arrange`. In this first mapping the 4 class approach is analyzed.

Considering the comparative mapping results of supervised classification algorithms with 4 classes of targets sampled in the landscape, apparently the vegetation target predominated in all classification situations performed. Given the generalization process, it is difficult to determine how much the fusion process interfered with the classification results, apparently determining more variability in the case of the `rpart1SE` algorithm under fusion compared to multispectral `rpart1SE`. In the case of the `svmLinear2` algorithm, fusion seems to have increased the generalization ability of classification results when compared to multispectral data. In the case of the `rf` algorithm there are doubts in visually determining differences between treatments with and without fusion (Figure 12.10).

12.2 Solved Exercises

12.2.1 Explain supervised classification of digital imagery.

A: Digital image classification is the process of assigning classes to pixels to determine similarity groups associated with the remotely sensed information of interest and generate a thematic digital map with satisfactory accuracy for a given purpose. Training samples in some regions of the image are used in a machine learning process to train computational intelligence algorithms to classify the entire image based on the patterns provided.

12.2.2 What are the advantages and disadvantages of supervised classification of digital imagery?

A: The advantages of supervised classification compared to the unsupervised method are: Control of the classes and the geographic region of interest in the classification; training data is known in the region; spectral mixing between classes can be minimized by choosing distinct training data

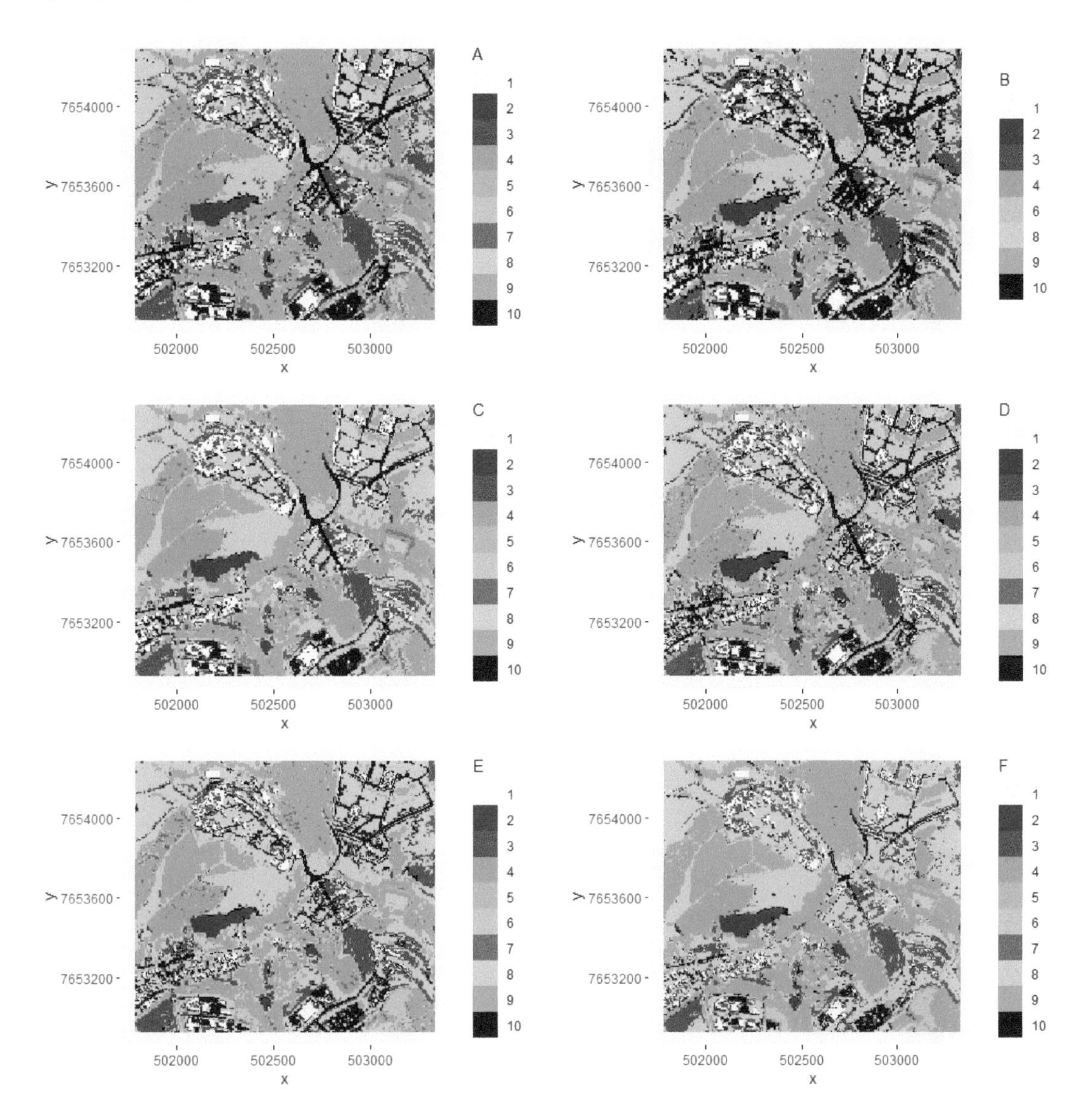

FIGURE 12.9 Mapping results of supervised classification with `rf` (A, D), `rpart1SE` (B, E), `svmLinear2` (C, D) algorithms considering 10 classes, without (A, B, C) and with (D, E, F) spectral band fusion, in the region of Lavras, state of Minas Gerais, on August 15, 2021.

features; the analyst can detect classification errors by examining the training data and the results of the classifier algorithm. The disadvantages of supervised classification are: The analyst imposes a classification structure on the data; the classes defined by the analyst may not be distinct in multidimensional space; training data are chosen first on the basis of classes and subsequently on spectral features; training data may not be representative according to the conditions observed in the entire image; the process of choosing training data can be time-consuming, expensive and tiring; there may be overlapping problems in choosing classes according to the spatial resolution of the image used; specific classes cannot be recognized in images if they are not included in the training data due to lack of knowledge or because they occupy small areas in the image.

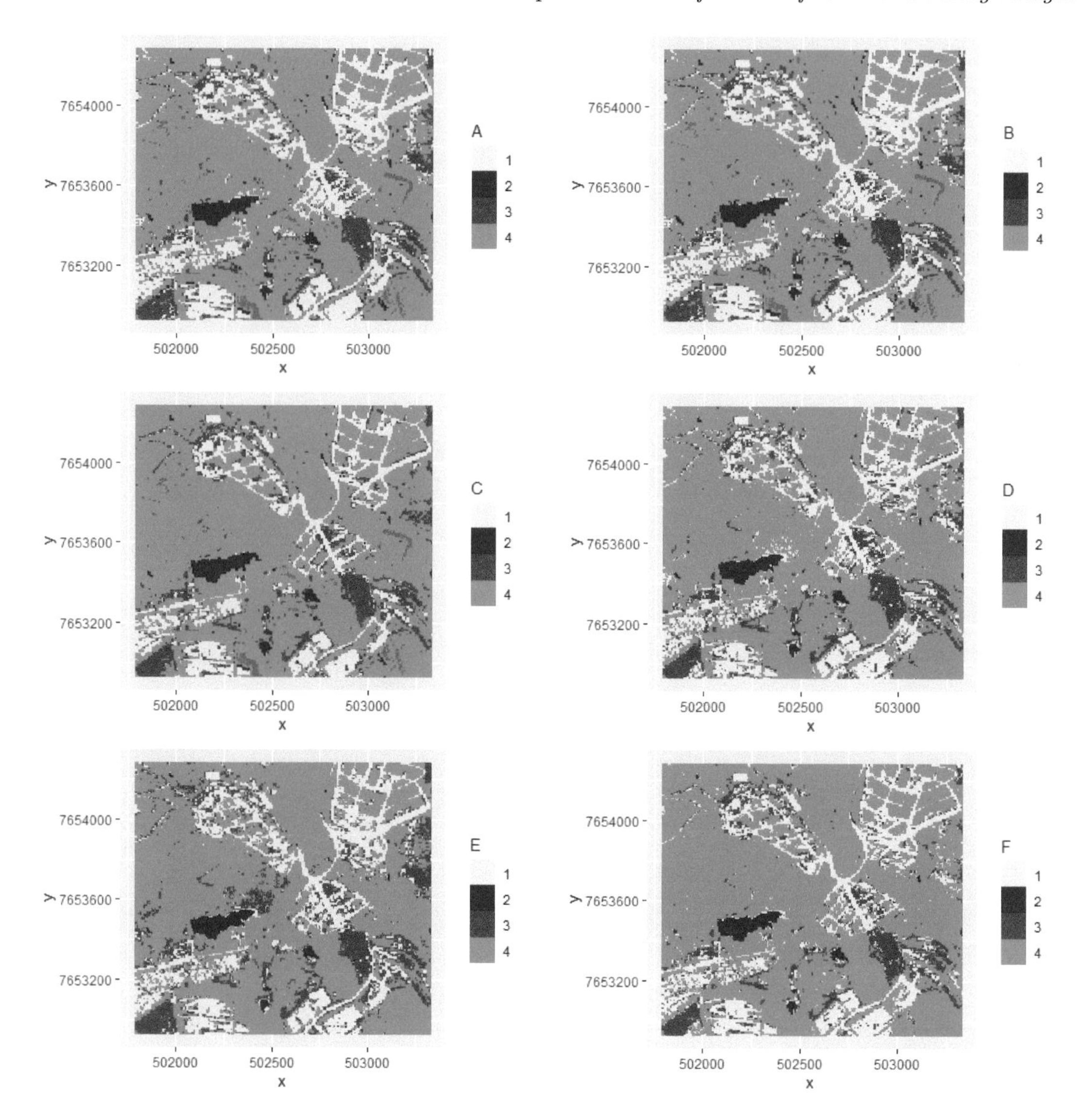

FIGURE 12.10 Mapping results of supervised classification with `rf` (A, D), `rpart1SE` (B, E), `svmLinear2` (C, D) algorithms considering 4 classes, without (A, B, C) and with (D, E, F) spectral band fusion, in the region of Lavras, state of Minas Gerais, on August 15, 2021.

12.2.3 Cite three algorithms for supervised classification of digital images.

A: Random forest, support vector machine, and neural networks.

12.2.4 Cite three application areas of supervised classification and the type of image used.

A: Land use classification with a Landsat-8 OLI image, urban area classification with a WorldView-2 image, and a drone image classification for mapping weeds in crops.

12.2.5 What are the important features for obtaining reliable training samples?

A: The important characteristics that should be considered when selecting training areas are: Number of pixels; size; shape; location and uniformity.

13

Uncertainty and Accuracy Analysis in Remote Sensing and Digital Image Processing

13.1 Homework Solution

13.1.1 Subject

Accuracy analysis of supervised classification applied to multispectral imagery from the CBERS-04A satellite for environmental monitoring.

13.1.2 Abstract

Confusion matrix analysis in remote sensing can be used to analyze the accuracy of supervised learning applied to image classification. The quality of spatial predictions of classification algorithms is evaluated from input data sets used in the training phase of algorithms, and also from validation by partitioning input data or using an additional database. We aimed to determine the confusion matrix and metrics of total accuracy and kappa index to evaluate the results of a remote sensing data analysis experiment with supervised classification of multispectral imagery from the CBERS-04A satellite, WPM camera, August 15, 2021. Spatial polygons are created under different arrangements of 10 and 4 classes to evaluate the effect of class generalization on algorithm training with and without panchromatic band fusion. In most of the analyzed situations the random forest classification algorithm performed better than `svmLinear2` and `rpart1SE` based on the accuracy analysis metrics. In both analyzed situations the accuracy analysis is higher than 95%, and it is necessary to carefully evaluate the logic of each classifier, and the advantages and disadvantages, for the scientific use of the classification results. Information about class mixing observed in the confusion matrix can be used as an experiment to improve future supervised classification trials in the monitored region. The result of the random forest classification under fusion and 4 classes is used as an example of segmentation for recognition of the classes in a vector attribute table considering the similarity criteria based on Euclidean, Jensen-Shannon, dynamic time warping and Manhattan distances.

Keywords: Accuracy analysis, classification algorithms, image fusion, kappa, machine learning, segmentation, WPM camera.

13.1.3 Introduction

In supervised remote sensing classification, questions may arise about which result to choose for scientific applications of results. The fundamental principle of accuracy analysis of a thematic classification result is determined as a function of the cross-comparison between the classification produced and the spatial reference considered to be true or observed. It is important that the reference used be representative in space and time of the same categorical universe. Statistical criteria and metrics can then be used to categorically compare the classification produced and the reference data. It is up to the analyst to carefully analyze the results of the confusion matrix in an attempt to select the best results in thematic applications or to re-adapt the classification scheme in the perspective of obtaining reliable results for scientific use. Selected results from pixel-by-pixel classification can be segmented into a hybrid classification approach (Moreira, 2011).

13.1.4 Objective

The objective of the homework is to perform confusion matrix analysis to determine classification quality metrics from an experiment used to evaluate different class arrangements with and without multispectral band fusion with the panchromatic imaging band of the CBERS-04A, WPM camera. Evaluate different distance criteria to determine pixel similarity in a hybrid classification approach using segmentation.

The R packages `RStoolbox` (Leutner et al., 2019), `caret` (Kuhn et al., 2020), `supercells` (Nowosad et al., 2022), `sf` (Pebesma et al., 2021), `terra` (Hijmans et al., 2022), and `raster` (Hijmans et al., 2020) are needed for analysis and they enabled with the `library` function.

```
library(RStoolbox)
library(caret)
library(supercells)
library(sf)
library(terra)
library(raster)
```

13.1.5 Pixel-by-pixel classification

The `readRSTBX` function is used to import the classification results into R.

```
SC1a <- readRSTBX("C:/sr/c10/new/SC1a")
SC2a <- readRSTBX("C:/sr/c10/new/SC2a")
SC3a <- readRSTBX("C:/sr/c10/new/SC3a")
SC1b <- readRSTBX("C:/sr/c10/new/SC1b")
SC2b <- readRSTBX("C:/sr/c10/new/SC2b")
SC3b <- readRSTBX("C:/sr/c10/new/SC3b")
SC1c <- readRSTBX("C:/sr/c10/new/SC1c")
SC2c <- readRSTBX("C:/sr/c10/new/SC2c")
SC3c <- readRSTBX("C:/sr/c10/new/SC3c")
SC1d <- readRSTBX("C:/sr/c10/new/SC1d")
SC2d <- readRSTBX("C:/sr/c10/new/SC2d")
SC3d <- readRSTBX("C:/sr/c10/new/SC3d")
```

The `list` function is used to create a list of the templates used. The `resamples` function from the `caret` package is used to resample the list of models. The `bwplot` function is used to represent the accuracy analysis and kappa index in graphical form.

```
# Create model list
model_list <- list(rf = SC1a$model, rpart1SE = SC2a$model,
                   svmLinear2 = SC3a$model)
model_list1 <- list(rf = SC1b$model, rpart1SE = SC2b$model,
                    svmLinear2 = SC3b$model)
model_list2 <- list(rf = SC1c$model, rpart1SE = SC2c$model,
                    svmLinear2 = SC3c$model)
model_list3 <- list(rf = SC1d$model, rpart1SE = SC2d$model,
                    svmLinear2 = SC3d$model)
# Resample model list
resamples <- caret::resamples(model_list)
resamples1 <- caret::resamples(model_list1)
resamples2 <- caret::resamples(model_list2)
resamples3 <- caret::resamples(model_list3)
# Display metrics in boxplots
a <- bwplot(resamples, main = "Multispectral (10 classes)")
b <- bwplot(resamples1, main = "Multispectral (4 classes)")
c <- bwplot(resamples2, main = "Fusion (10 classes)")
d <- bwplot(resamples3, main = "Fusion (4 classes)")
grid.arrange(a,b,c,d, ncol=1)
```

In the model fitting phase, the higher accuracy and kappa values were found for the `rf` algorithm, followed by `svmLinear2` and `rpart1SE`. However, all algorithms used showed satisfactory performance to classify the targets, and it is important to evaluate other features of each model to perform the classification for scientific use of results (Figure 13.1)

The confusion matrix is obtained for each of the evaluated treatments `rf`, `rpart1SE`, and `svmLinear2` Multispectral 10 classes; `rf`, `rpart1SE`, and `svmLinear2` Multispectral 4 classes; `rf`, `rpart1SE`, and `svmLinear2` Fusion 10 classes; `rf`, `rpart1SE`, and `svmLinear2` Fusion 4 classes.

```
SC1a$modelFit # rf
#Cross-Validated (10 fold) Confusion Matrix
#(entries are average cell counts across resamples)
#           Reference
#Prediction     1     2     3     4     5     6     7     8     9    10
#         1  17.9   0.0   0.0   0.0   0.0   0.0   0.0   0.0   0.0   0.0
#         2   0.0 100.1   0.0   0.0   0.0   0.0   0.0   0.0   0.0   0.0
#         3   0.0   0.0  39.6   0.0   0.0   0.0   0.0   0.0   0.0   0.0
#         4   0.0   0.0   0.0  99.3   0.0   0.0   0.2   0.0   0.0   0.0
#         5   0.0   0.0   0.0   0.0   2.3   0.0   0.1   0.0   0.0   0.0
#         6   0.0   0.0   0.0   0.0   0.0  33.8   0.0   0.0   0.0   0.0
#         7   0.0   0.0   0.0   0.0   0.0   0.0   5.1   0.0   0.0   0.0
#         8   0.0   0.0   0.0   0.0   0.0   0.0   0.0  58.8   0.0   0.0
#         9   0.0   0.0   0.0   0.0   0.2   0.0   0.2   0.0  96.0   0.0
#        10   0.0   0.0   0.0   0.0   0.0   0.0   0.0   0.0   0.0  17.7
# Accuracy (average) : 0.9985
SC2a$modelFit # rpart1SE
#Cross-Validated (10 fold) Confusion Matrix
#(entries are average cell counts across resamples)
#           Reference
#Prediction    1    2    3    4    5    6    7    8    9   10
```

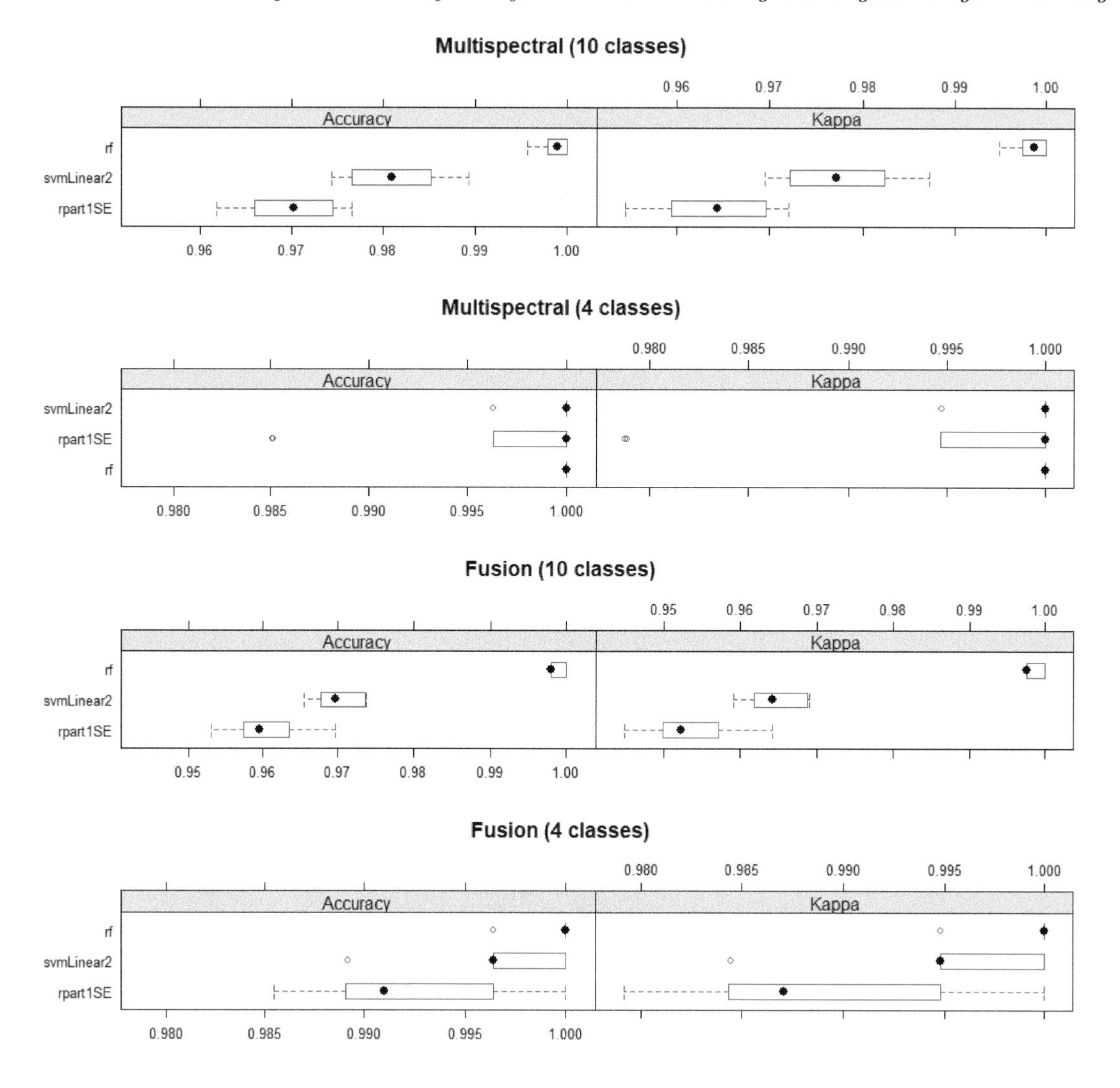

FIGURE 13.1 Accuracy analysis and kappa index in model fitting after supervised classification of remote sensing big data to compare `rf`, `svmLinear2`, and `rpart1SE` algorithms considering 10 and 4 training classes, and image processing with and without spectral band fusion, in the region of Lavras, state of Minas Gerais, on August 15, 2021.

```
#        1  17.9  0.0  0.0  0.0  0.0  0.0  0.0  0.0  0.0  0.0
#        2   0.0 98.5  0.0  0.0  0.0  0.0  0.0  0.0  0.0  0.0
#        3   0.0  0.0 39.6  0.0  0.0  0.0  0.0  0.0  0.0  0.0
#        4   0.0  0.0  0.0 96.8  0.7  0.5  4.7  0.0  1.6  0.0
#        5   0.0  0.0  0.0  0.0  0.0  0.0  0.0  0.0  0.0  0.0
#        6   0.0  0.0  0.0  0.0  1.6 33.3  0.7  0.0  0.0  0.0
#        7   0.0  0.0  0.0  0.0  0.0  0.0  0.0  0.0  0.0  0.0
#        8   0.0  0.0  0.0  0.0  0.0  0.0  0.0 58.8  0.0  0.0
#        9   0.0  0.0  0.0  2.4  0.2  0.0  0.2  0.0 94.4  0.0
#       10   0.0  1.6  0.0  0.0  0.0  0.0  0.0  0.0  0.0 17.7
# Accuracy (average) : 0.9699
SC3a$modelFit # svmLinear2
```

```
#Cross-Validated (10 fold) Confusion Matrix
#(entries are average cell counts across resamples)
#          Reference
#Prediction     1     2     3     4     5     6     7     8     9    10
#         1  17.9   0.0   0.0   0.0   0.0   0.0   0.0   0.0   0.0   0.0
#         2   0.0 100.0   0.0   0.0   0.0   0.0   0.0   0.0   0.0   0.0
#         3   0.0   0.0  39.6   0.0   0.0   0.0   0.0   0.0   0.0   0.0
#         4   0.0   0.0   0.0  99.4   0.0   0.0   5.3   0.0   3.2   0.0
#         5   0.0   0.0   0.0   0.0   2.5   0.0   0.1   0.0   0.0   0.0
#         6   0.0   0.0   0.0   0.0   0.0  33.8   0.0   0.0   0.0   0.0
#         7   0.0   0.0   0.0   0.0   0.0   0.0   0.0   0.0   0.0   0.0
#         8   0.0   0.1   0.0   0.0   0.0   0.0   0.0  58.8   0.0   0.0
#         9   0.0   0.0   0.0   0.0   0.0   0.0   0.2   0.0  92.8   0.0
#        10   0.0   0.0   0.0   0.0   0.0   0.0   0.0   0.0   0.0  17.7
# Accuracy (average) : 0.9811
SC1b$modelFit # rf
#Cross-Validated (10 fold) Confusion Matrix
#(entries are average cell counts across resamples)
#          Reference
#Prediction     1     2     3     4
#         1  35.6   0.0   0.0   0.0
#         2   0.0 100.1   0.0   0.0
#         3   0.0   0.0  39.6   0.0
#         4   0.0   0.0   0.0  93.7
# Accuracy (average) : 1
SC2b$modelFit # rpart1SE
#Cross-Validated (10 fold) Confusion Matrix
#(entries are average cell counts across resamples)
#          Reference
#Prediction    1    2    3    4
#         1 35.6  0.8  0.0  0.1
#         2  0.0 99.3  0.0  0.0
#         3  0.0  0.0 39.6  0.0
#         4  0.0  0.0  0.0 93.9
# Accuracy (average) : 0.9967
SC3b$modelFit # svmLinear2
#Cross-Validated (10 fold) Confusion Matrix
#(entries are average cell counts across resamples)
#          Reference
#Prediction     1     2     3     4
#         1  35.6   0.0   0.0   0.0
#         2   0.0 100.0   0.0   0.0
#         3   0.0   0.0  39.6   0.0
#         4   0.0   0.1   0.0  94.0
# Accuracy (average) : 0.9996
SC1c$modelFit # rf
#Cross-Validated (10 fold) Confusion Matrix
#(entries are average cell counts across resamples)
#          Reference
#Prediction     1     2     3     4     5     6     7     8     9    10
#         1  17.9   0.0   0.0   0.0   0.0   0.0   0.0   0.0   0.0   0.0
#         2   0.0 100.1   0.0   0.0   0.0   0.0   0.0   0.0   0.0   0.0
```

```
#        3    0.0    0.0  39.6    0.0    0.0    0.0    0.0    0.0    0.0    0.0
#        4    0.0    0.0    0.0 100.2    0.0    0.0    0.6    0.0    0.0    0.0
#        5    0.0    0.0    0.0    0.0  18.3    0.0    0.0    0.0    0.0    0.0
#        6    0.0    0.0    0.0    0.0    0.0  35.7    0.0    0.0    0.0    0.0
#        7    0.0    0.0    0.0    0.0    0.0    0.0    9.1    0.0    0.0    0.0
#        8    0.0    0.0    0.0    0.0    0.0    0.0    0.0  58.8    0.0    0.0
#        9    0.0    0.0    0.0    0.0    0.0    0.0    0.0    0.0  96.0    0.0
#       10    0.0    0.0    0.0    0.0    0.0    0.0    0.0    0.0    0.0  17.7
# Accuracy (average) : 0.9988
SC2c$modelFit # rpart1SE
#Cross-Validated (10 fold) Confusion Matrix
#(entries are average cell counts across resamples)
#          Reference
#Prediction    1    2    3    4    5    6    7    8    9   10
#        1  16.8  0.0  0.0  0.0  0.0  0.0  0.0  0.0  0.0  0.0
#        2   0.0 98.3  0.0  0.0  0.0  0.0  0.0  0.0  0.0  0.0
#        3   0.0  0.0 39.6  0.0  0.0  0.0  0.0  0.2  0.0  0.0
#        4   0.0  0.2  0.0 93.4  0.9  0.0  5.2  0.0  2.2  0.0
#        5   0.0  0.0  0.0  0.0 16.6  0.0  0.0  0.0  0.0  0.0
#        6   0.0  0.0  0.0  0.0  0.1 35.7  0.0  0.0  0.0  0.0
#        7   0.0  0.0  0.0  4.3  0.0  0.0  4.3  0.0  0.0  0.0
#        8   1.0  0.0  0.0  0.0  0.0  0.0  0.0 58.6  0.0  0.4
#        9   0.0  0.0  0.0  2.5  0.7  0.0  0.2  0.0 93.8  0.0
#       10   0.1  1.6  0.0  0.0  0.0  0.0  0.0  0.0  0.0 17.3
# Accuracy (average) : 0.9603
SC3c$modelFit # svmLinear2
#Cross-Validated (10 fold) Confusion Matrix
#(entries are average cell counts across resamples)
#          Reference
#Prediction    1    2    3    4    5    6    7    8    9   10
#        1  17.7  0.0  0.0  0.0  0.0  0.0  0.0  0.0  0.0  0.0
#        2   0.0 99.8  0.0  0.0  0.0  0.0  0.0  0.0  0.0  0.5
#        3   0.2  0.0 39.6  0.0  0.0  0.0  0.0  0.0  0.0  0.0
#        4   0.0  0.0  0.0 97.1  0.0  0.0  8.3  0.0  1.6  0.0
#        5   0.0  0.0  0.0  0.0 18.3  0.0  0.0  0.0  0.0  0.0
#        6   0.0  0.0  0.0  0.0  0.0 35.7  0.6  0.0  0.0  0.0
#        7   0.0  0.0  0.0  0.5  0.0  0.0  0.6  0.0  0.0  0.0
#        8   0.0  0.0  0.0  0.0  0.0  0.0  0.0 58.8  0.0  0.0
#        9   0.0  0.0  0.0  2.6  0.0  0.0  0.2  0.0 94.4  0.0
#       10   0.0  0.3  0.0  0.0  0.0  0.0  0.0  0.0  0.0 17.2
# Accuracy (average) : 0.97
SC1d$modelFit # rf
#Cross-Validated (10 fold) Confusion Matrix
#(entries are average cell counts across resamples)
#          Reference
#Prediction     1     2     3     4
#          1  35.6   0.1   0.1   0.0
#          2   0.0 100.0   0.0   0.0
#          3   0.0   0.0  39.5   0.0
#          4   0.0   0.0   0.0 100.9
# Accuracy (average) : 0.9993
SC2d$modelFit # rpart1SE
```

```
#Cross-Validated (10 fold) Confusion Matrix
#(entries are average cell counts across resamples)
#           Reference
#Prediction     1     2     3     4
#          1  34.6   1.1   0.0   0.0
#          2   0.0  99.0   0.0   0.0
#          3   0.0   0.0  39.6   0.0
#          4   1.0   0.0   0.0 100.9
# Accuracy (average) : 0.9924
SC3d$modelFit # svmLinear2
#Cross-Validated (10 fold) Confusion Matrix
#(entries are average cell counts across resamples)
#           Reference
#Prediction     1     2     3     4
#          1  35.0   0.3   0.0   0.0
#          2   0.4  99.8   0.0   0.0
#          3   0.2   0.0  39.6   0.0
#          4   0.0   0.0   0.0 100.9
# Accuracy (average) : 0.9967
```

13.1.6 Segmentation

The random forest classification result under fusion and 4 classes is exported to a directory of interest with the `writeRaster` function.

```
writeRaster(SC1b$map, "C:/sr/c10/new/rasterSC1b.tif", overwrite=TRUE)
```

The classified raster layer is imported into R with the `rast` function.

```
landcover <- rast("C:/sr/c10/new/rasterSC1b.tif")
```

The `supercells` function is used in the segmentation of the pixel-by-pixel classification by the Euclidean distance, Jensen-Shannon, dynamic time warping, and Manhattan methods.

```
landCell1 <- supercells(landcover, step = 20, compactness = 1,
           dist_fun = "euclidean", clean = FALSE, iter = 10)
landCell2 <- supercells(landcover, step = 20, compactness = 1,
           dist_fun = "jsd", clean = FALSE, iter = 10) #Jensen-Shannon
landCell3 <- supercells(landcover, step = 20, compactness = 1,
           dist_fun = "dtw", clean = FALSE, iter = 10) #dynamic time warping
landCell4 <- supercells(landcover, step = 20, compactness = 1,
           dist_fun = "manhattan", clean = FALSE, iter = 10) #Manhattan
```

The segmentation results are mapped comparatively with the `par(mfrow)` and `plot` functions.

```
par(mfrow=c(2,2))
LCcolors <- c("grey90","cyan","red","green4")
```

```
plot(landcover, col=LCcolors, main = "Euclidean")
plot(st_geometry(landCell1), add = TRUE, lwd = 0.2)
plot(landcover, col=LCcolors, main = "Jensen-Shannon")
plot(st_geometry(landCell2), add = TRUE, lwd = 0.2)
plot(landcover, col=LCcolors, main = "Dynamic time warping")
plot(st_geometry(landCell3), add = TRUE, lwd = 0.2)
plot(landcover, col=LCcolors, main = "Manhattan")
plot(st_geometry(landCell4), add = TRUE, lwd = 0.2)
```

Regardless of the distance method evaluated, the segmentation results are visually similar (Figure 13.2), although the Jensen-Shannon distance determined the smallest storage size when compared to the other methods.

```
object.size(landCell1)
#2321520 bytes
object.size(landCell2)
#1761424 bytes
object.size(landCell3)
#2321520 bytes
object.size(landCell4)
#2321520 bytes
```

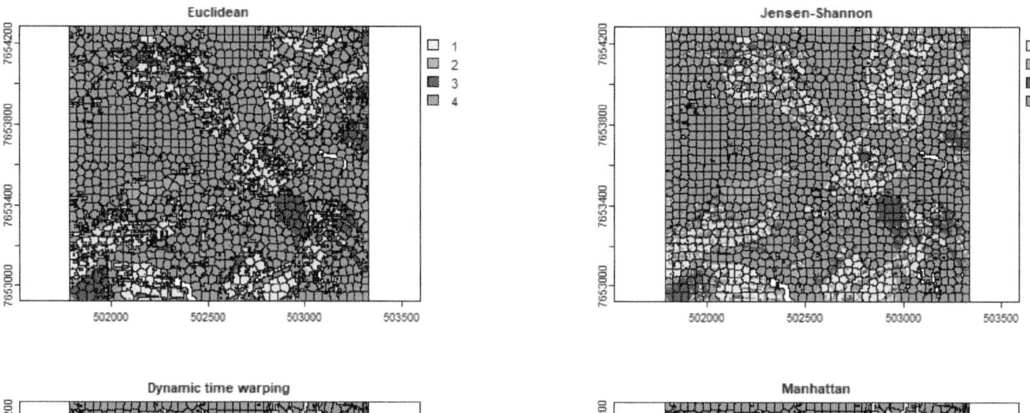

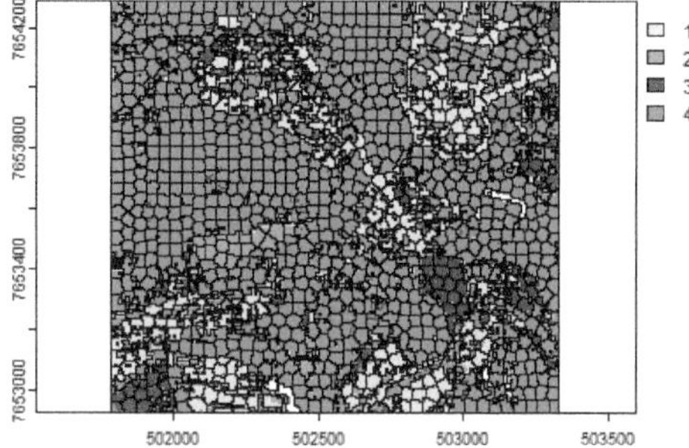

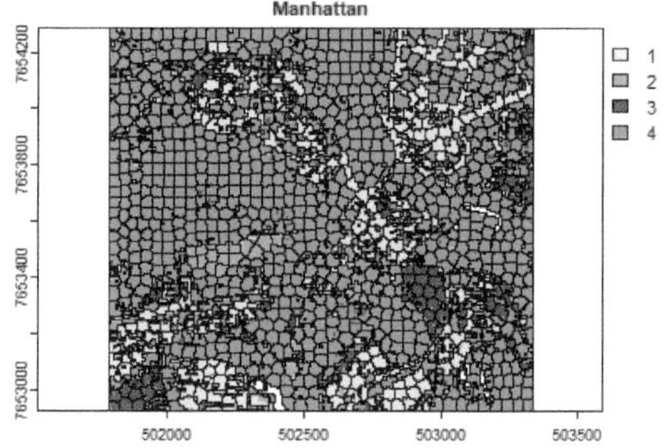

FIGURE 13.2 Supercells segmentation of pixel-by-pixel classification results in the region of Lavras, state of Minas Gerais, on August 15, 2021.

13.2 Solved Exercises

13.2.1 What are the sources of errors in information obtained by remote sensing?

A: The sources of error in thematic products obtained by remote sensing varied according to sensor limitation, analysis method, and target complexity from the data and image acquisition process by passive and active remote sensing systems, through preprocessing and modeling steps, to the final product.

13.2.2 What are the steps to determine classification accuracy?

A: The steps to obtain accuracy of thematic information from remote sensing data are: Determine the nature of the thematic information to be evaluated, define methods for evaluating thematic accuracy, determine the total number of observations needed in the analysis, define the sampling scheme; obtain reference information on the ground; perform the error matrix, and accept or reject the hypothesis according to the acceptance results of the accuracy magnitude.

13.2.3 Determine the number of points required to evaluate a classification with expected accuracy of 85% and tolerable error of 5% based on binomial probability theory.

A:

```
# N=2^2(85)(15)/5^2
N<-2^2*85*15/5^2
N
```

```
## [1] 204
```

```
#[1] 204
```

13.2.4 Determine the number of points required to evaluate a classification with expected accuracy of 85% and tolerable error of 10% based on binomial probability theory.

A:

```
# N=2^2(85)(15)/10^2
N<-2^2*85*15/10^2
N
```

```
## [1] 51
```

```
#[1] 51
```

13.2.5 Cite five types of sampling plans used in remote sensing map accuracy analysis.

A: Random, systematic, stratified random, unaligned stratified systematic, and clustered sampling.

13.2.6 What are the commands used to determine the silhouette index and confusion matrix in practice in R?

A: Silhouette index: `intCriteria`; confusion matrix: `getValidation`.

13.2.7 Evaluate the overall accuracy of a supervised classification according to the confusion matrix data of vegetation, soil, water, and urban targets (Table 13.1).

TABLE 13.1 Confusion matrix of the observed and predicted vegetation, soil, water, and urban targets by the naive Bayes classifier.

		Observed				
	Class	Vegetation	Soil	Water	Urban	Line total
	Vegetation	98	5	0	1	104
	Soil	0	34	36	4	74
Predicted	water	1	0	70	0	71
	Urban	0	26	0	129	155
	Column total	99	65	106	134	404

```
TotalLineVeg<-98+5+0+1
TotalLineVeg
#[1] 104
TotalLineSoil<-0+34+36+4
TotalLineSoil
#[1] 74
TotalLineWater<-1+0+70+0
TotalLineWater
#[1] 71
TotalLinhaUrban<-0+26+0+129
TotalLineUrban
#[1] 155
TotalColumnVeg<-98+0+1+0
TotalColumnVeg
#[1] 99
TotalColumnSoil<-5+34+0+26
```

```
TotalColumnSoil
#[1] 65
TotalColumnWater<-0+36+70+0
TotalColumnWater
#[1] 106
TotalColumnUrban<-1+4+0+129
TotalColumnUrban
#[1] 134
N<-104+74+71+155 # sum of row totals
N
#[1] 404
N<-99+65+106+134 # sum of total columns (check)
N
#[1] 404
# Determine the number of correct matches (sum of the diagonal of the table)
hits<-98+34+70+129
hits
#[1] 331
# Determine the total accuracy
total_accuracy<-(hits/N)*100
total_accuracy
#[1] 81.93069
# Determine the product of row summation by column
prodTotalRowColumn <- ((TotalRowVeg*TotalColumnVeg)+
    (TotalRowSoil*TotalColumnSoil)+ (TotalRowWater*TotalColumnWater)+
      (TotalRowUrban*TotalColumnUrban))
prodTotalLineColumn
#[1] 43402
# Determine Kappa
Kappa <-((N*hits)-prodTotalRowColumn)/((N^2)-prodTotalRowColumn)*100
Kappa
#[1] 75.38518
```

A: The overall accuracy is 81.93%; the kappa is 75.38%.

13.2.8 In a classification analysis of remote sensing data in which the class `water` (1) and `non-water` (0) are evaluated in terms of prediction of the classifier algorithm, determine the confusion matrix, accuracy, sensitivity and specificity of the classification, based on the following data:

Observed pixel values with water and non-water: 1, 1, 0, 1, 0, 0, 1, 0, 0, 0.

Predicted pixel values with water and non-water: 1, 0, 0, 1, 0, 0, 1, 1, 0.

A: The `caret` (Kuhn et al., 2020) and `e1071` (Meyer et al., 2021) packages installed in R are enabled to solve the exercise.

```
library(caret)
library(e1071)
```

The `factor` function is used to record the observed and predicted data as `Ob` and `Pr`, respectively.

```
Ob<-factor(c(1, 1, 0, 1, 0, 0, 1, 0, 0, 0))
Pr<-factor(c(1, 0, 0, 1, 0, 0, 1, 1, 1, 0))
```

The `confusionMatrix` function is used to generate the results of the accuracy analysis.

```
res<-confusionMatrix(data=Pr, reference=Ob)
res
#Confusion Matrix and Statistics
#          Reference
#Prediction 0 1
#         0 4 1
#         1 2 3
#               Accuracy : 0.7
#                 95% CI : (0.3475, 0.9333)
#    No Information Rate : 0.6
#    P-Value [Acc > NIR] : 0.3823
#                  Kappa : 0.4
# Mcnemar's Test P-Value : 1.0000
#            Sensitivity : 0.6667
#            Specificity : 0.7500
#         Pos Pred Value : 0.8000
#         Neg Pred Value : 0.6000
#             Prevalence : 0.6000
#         Detection Rate : 0.4000
#   Detection Prevalence : 0.5000
#      Balanced Accuracy : 0.7083
#       'Positive' Class : 0
```

After determining the confusion matrix, the classification accuracy, sensitivity, and specificity are 0.70, 0.66, and 0.75, respectively.

14

Remote Sensing and Digital Image Processing for Article Enhancement

14.1 Homework Solution

14.1.1 Subject

Multisensor analysis for identification of coffee crops and other environmental targets in the Funil dam region, state of Minas Gerais, Brazil.

14.1.2 Abstract

Remote sensing has been viable for scalable, replicable and affordable mapping of land cover and land use change. Accurate details about the landscape are provided by classifying satellite images through image processing techniques, modeling, and use of classification algorithms. We aimed to evaluate a multisensor approach to identify coffee plantations, water, urban landscape, forest, bare soil, and grassland landscape in the Funil dam region, state of Minas Gerais, Brazil. Landsat-8, Sentinel-1, Sentinel-2 imagery data were referenced to June 2021. Multisensor remote sensing big data was determined for target identification with `rf`, `rpart1SE` and `svmLinear2` supervised classification algorithms. Considering the classification of coffee areas specifically, the `svmLinear2` algorithm may be indicated, in view of the greater number of identified areas and the better distinction of vegetation areas. The Landsat-8 imagery data were preferred by the `rf` algorithm; however, in the `rpart1SE` algorithm, the Sentinel-2 band 8A (20 m) was fundamental in separating water from the other targets analyzed by establishing a threshold of 0.1031 in the first decision node. The analyzed classifiers were satisfactory to classify landscape targets in the Funil dam region, Minas Gerais, Brazil.

Keywords: Machine learning, supervised classification, Landsat-8, Sentinel-1, Sentinel-2.

14.1.3 Introduction

Remote sensing has been an increasingly effective tool for crop mapping (Gallego et al., 2014; Thenkabail & Wu, 2012; Wu et al., 2014). It represents the most feasible approach for scalable, replicable, and affordable mapping of land cover and land use change (Hunt et al., 2020). Among the methods used, manual classification based on visual interpretation and digital per-pixel classification are the most applied (Büttner, 2014; Tian et al., 2016). Visual interpretation, however, is considered subjective, time-consuming, and costly (Xiong et al., 2017). The availability of satellites allow gathering of high-resolution data for cropland mapping (Drusch et al., 2012). Accurate details about the landscape are provided by classifying satellite images through image processing techniques, modeling, and use of classification algorithms (Al-Ahmadi & Al-Hames, 2009).

The US Geological Survey's Landsat data are the most accessible for land cover mapping, due to the sensor's specific design for natural resource mapping, the longevity of the mission relative to other satellites, and the free release of the imagery. Sentinel-2 is considered more limited than Landsat when it comes to spectral pixel-based coffee mapping due to its launch only in 2015 (Hunt et al., 2020). However, Sentinel-2 is a viable data source for mapping as it overcomes cloud cover issues (Claverie et al., 2018). In addition, its 10-, 20-, and 30-m resolution ranges, 5-day revisit time, and free image availability are factors responsible for its potential in plant species mapping (Hunt et al., 2020). Another tool for monitoring the Earth's surface is Synthetic Aperture Radar (SAR), which is developing rapidly with the recent launches of several space satellites, such as Sentinel-1, Chinese Gaofen-3, and India Risat-1. Such satellites are provided with various imagery modes, which broaden the application scope and information content for different demands. Using multi-temporal or multi-polarimetric observations, croplands can be well characterized through classification (Larrañaga & Álvarez-Mozos, 2016).

Given the complexity and variety of coffee production systems, there is still some limitation on mapping this crop (Hunt et al., 2020). Some factors, such as cultivation in small plots, usually in heterogeneous landscapes, at higher elevations, i.e., with steep and complex topographies, cause problems with image interpretation (Bernardes et al., 2012; Cordero-Sancho & Sader, 2007; Langford & Bell, 1997; Lu et al., 2008). Because it is spectrally similar to various woody cover types, it can then be easily confused with other crops, vegetation, and other land cover types during classification approaches (Bolanos, 2007; Langford & Bell, 1997; Schmitt-Harsh, 2013).

Other factors to consider are the spectral signature of coffee, which varies with age and over its two-year phenological cycle, as well as during rust outbreaks (Bernardes et al., 2012; Kushalappa & Eskes, 1989). Although such features reduce the accuracy of coffee mapping using satellite data, Bourgoin et al. (2020) used Sentinel-2 to map land cover in Vietnam, including coffee extensions, generating a landscape analysis and ecological vulnerability assessment. Belgiu & Csillik (2018) verified the utility of using object-based image analysis approaches on Sentinel-2 data in monitoring agricultural growing regions. Schmitt-Harsh (2013) used a pixel-based maximum likelihood classifier on Landsat imagery and spectral mixture analysis identifying shadow, soil, and vegetation. By incorporating the images into the classification, including optical and thermal bands, there was an accuracy of 88.6% and a user accuracy of 89.7% for coffee agroforestry monitoring. Spectral mixture analysis can be effective because it considers that physical processes cause the observed spectral signature and therefore incorporates mixed pixels (Kawakubo & Pérez Machado, 2016). Sentinel-1 RADAR data also offer an excellent option for mapping coffee because they allow a texture-based approach that permits capturing aspects of canopy structure, a key feature for distinguishing coffee agroecosystems (Hunt et al., 2020). Fusion of optical aperture radar and synthetic radar (specifically Sentinel-1 and Sentinel-2 data) has been applied in coffee mapping (Hunt et al., 2020). The inclusion of Sentinel-1 has the main advantage of not being restricted by cloud cover. There are a growing number of studies integrating these two data types in crop mapping (Qadir & Mondal, 2020).

14.1.4 Objective

The objective of this work was to evaluate the importance of variables in the identification of coffee crops, water, urban landscape, forest, bare soil, and grassland landscape using a multisensor approach in the area of Funil dam, Minas Gerais. The performance of different algorithms was analyzed for predicting landscape targets at a given time and identifying the spectral signature with exploratory data analysis.

14.1.5 Material and methods

14.1.5.1 Location and description of the study area

The Funil hydroelectric plant is located in the Alto Rio Grande plateau, in the Rio Grande sub-basin, belonging to the Paraná River basin, between the municipalities of Perdões and Lavras, Minas Gerais, upstream of the Furnas UHE reservoir and downstream of Itutinga UHE and Camargos UHE. The area is delimited by the geographic coordinates 21° 06'24" and 21° 13'60" S latitude and 45° 04'38" and 44° 54'13" W longitude.

The vegetation in the region is formed by the transition between cerrado and semideciduous forests. Rupestrian fields and altitude fields are identified, located over shallow and young soils in high-altitude areas (Oliveira Filho et al., 1994). Regarding geology, the basement is composed of quartzite, mica schists, leucocratic and mesocratic granitic gneisses. There are limestones in the municipality of Ijaci and mostly sandy-siltstone sediments near the Grande River. According to the Köppen climate classification, the climate in the region is Cwb type, with irregular rainfall distributed throughout the year, with greater volume in the months of November through March and deficiency between April and August (Dantas et al., 2007). During the summer the maximum annual pluviometric index is approximately 2500 mm, while in the winter, the annual index is 500 mm, reaching an annual average of 1500 mm. The average annual temperature is 21°C, with an average temperature of 24°C, in the summer and an average of 7°C, in the coldest month of the year, July (Messias & Ferreira, 2014).

14.1.5.2 Satellite data acquisition

Sentinel-1 and Sentinel-2 imagery data were acquired via the European Space Agency (ESA) Portal and data from the Landsat-8 satellite via the United States Geological Survey (USGS), Earth Explorer portal. The Landsat-8 OLI/TIRS imagery data were taken on June 28, 2021, Sentinel-1 GRD and Sentinel-2 on June 25, 2021 (Table 14.1). The data were geometrically corrected and calibrated to the WGS-84 reference point and projected into the UTM system.

TABLE 14.1 Information about the remote sensing data used.

Sensor	Date	Source	Format	Reference Systems	Spatial Res. (m)
Landsat-8	June 28, 2021	NASA	raster	WGS-84	30
Sentinel-1	June 25, 2021	ESA	raster	WGS-84	10-30
Sentinel-2	June 25, 2021	ESA	raster	WGS-84	10-30

ESA has developed a number of satellites with great ability to obtain physical, biological, and biophysical information from the Earth's surface (Malenovský et al., 2012). The Sentinel-1 mission, for example, is seen with great potential in operational SAR missions in the coming decades (Krassenburg, 2016). The Sentinel-1 system is based on a constellation of two SAR satellites, Sentinel-1A, launched April 3, 2014 and Sentinel-1B, launched April 25, 2016, both with integrated C-band sensors. With a 12-day revisit time, Sentinel-1 operates in four unique acquisition modes: Stripmap (SM), Interferometric Wide Swath (IW), Extra Wide Swath (EW) and Wave (WV). Sentinel-1 ensures data availability through global coverage and day and night data for all weather conditions (Washaya et al., 2018).

The S2 satellite has a near-polar orbit and is equipped with an MSI (multispectral instrument) sensor, which has a width of 290 km and dimension of 100 by 100 km, 12 bits per pixel, with the ability to obtain information from 13 spectral bands ranging from the visible, near-infrared, and shortwave infrared. Spatial resolution ranges from 10 to 60 m and temporal resolution of 5 days with both satellites operational.

Landsat-8 was launched on February 11, 2013 by the National Aeronautics and Space Administration (NASA), with the Thermal Infrared Sensor (TIRS) and Operational Land Imager (OLI) sensors that have provided calibrated high spatial resolution data of the earth's surface for over 40 years (Matias, 2019). The L8 also features two new spectral bands, band 1 corresponding to coastal aerosol and band 9 to cirrus. The TIRS sensor has a new thermal infrared band with about 30 m of spatial resolution. The OLI sensor has 9 shortwave spectral bands, a scanning range of 190 km, and a spatial resolution of 30 meters for all bands except band 8 (panchromatic) with 15 m (Matias, 2019).

14.1.5.3 Data analysis

A multisensor remote sensing big data was determined for identification of coffee crops and other environmental targets in the Funil dam area, state of Minas Gerais, Brazil. The input variables considered in the classification analysis are listed with the raster layer name and the source sensor (Table 14.2).

TABLE 14.2 Raster layers and source sensor used as input data in supervised classification.

Source Sensor	Raster Layers
Landsat-8	"costal", "blue", "green", "red", "nir", "swir2", "swir3", "swir1", "tir1", "tir2", "brightness", "greenness", "wetness", "CTVI.1", "DVI.1", "EVI.1", "EVI2", "GEMI.1", "GNDVI.1", "MNDWI.1", "MSAVI.1", "MSAVI2.1", "NBRI.1", "NDVI.1", "NDWI.1", "NDWI2.1", "NRVI.1", "RVI.1", "SATVI.1", "SAVI.1"
Sentinel-1	"S1B_IW_GRDH_1SDV_20210628T082946_Cal_Spk_TC.3", "S1B_IW_GRDH_1SDV_20210628T082946_Cal_Spk_TC.4"
Sentinel-2	"T23KNS_20210625T131249_B02_10m", "T23KNS_20210625T131249_B03_10m", "T23KNS_20210625T131249_B04_10m", "T23KNS_20210625T131249_B05_20m", "T23KNS_20210625T131249_B06_20m", "T23KNS_20210625T131249_B07_20m", "T23KNS_20210625T131249_B08_10m", "T23KNS_20210625T131249_B8A_20m", "T23KNS_20210625T131249_B09_60m", "T23KNS_20210625T131249_B11_20m", "T23KNS_20210625T131249_B12_20m", "CLG", "CLRE", "CTVI.2", "DVI.2", "EVI.2", "GEMI.2", "GNDVI.2", "MCARI", "MNDWI.2", "MSAVI.2", "MSAVI2.2", "NBRI.2", "NDREI1", "NDREI2", "NDVI.2", "NDWI.2", "NDWI2.2", "NRVI.2", "RVI.2", "SATVI.2", "SAVI.2", "SLAVI", "SR", "WDVI"

More details on how these variables were preprocessed can be accessed in the book *Remote Sensing and Digital Image Processing with R.*

14.1.5.4 Obtaining training samples

The supervised classification process basically consists of three main steps, selection of training areas, image classification, and evaluation of results. Data samples were extracted from the images to evaluate the spectral signature of the targets. To collect the training samples, spatial polygons of the geographical regions of interest were made. As many repetitions of homogeneous and irregular polygon samples as possible were made for each identified target, with 463, 54, 53, 71, 60, and 51 features in polygon geometry of coffee crops, water, urban area, tree vegetation, bare ground, and grassland, respectively (Figure 14.1).

FIGURE 14.1 Interactive mapping of regions of interest on coffee crop targets (1), water (2), urban landscape (3), forest (4), bare soil (5), and grassland (6) to sample pixels per training class to perform supervised classification at Funil dam, state of Minas Gerais, Brazil.

14.1.5.5 Exploratory data analysis

A barplot was performed for exploratory analysis of mean values of spectral variables from Landsat-8 OLI sampled on targets related to coffee crops (1), water (2), urban landscape (3), forest (4), bare soil (5), grassland landscape (6) in the Funil dam region, Minas Gerais, Brazil.

The spectral signature of variables was also performed with a continuous two-dimensional plot of the variation of each band used in the analysis according to the reflective, thermal infrared and RADAR range spectra on the y axis and the wavelength on the x axis for each target analyzed with the Sentinel-1,2 and Landsat-8 satellite sensors.

14.1.5.6 Supervised classification

Pixel-by-pixel classification is automatic, given the spectral behavior that is the main element of the classifier. This classification uses isolated spectral attributes of each pixel in the image, assigning to each the most appropriate class. Pixel-by-pixel image classification techniques are divided into unsupervised and supervised classification. Supervised classifiers use algorithms to classify the pixels of an image into classes with different types of land use and land cover.

Supervised classification training was performed using Classification and Regression Trees (`CART`) for machine learning, employing the `rpart1SE` method. Other machine learning algorithms evaluated were Random Forest (`RF`) and Support Vector Machine (`SVM`) with the linear kernel (`svmLinear2`).

The `CART` classifier is a binary decision tree constructed by splitting a node in two, starting with the root node that contains the entire learning sample.

The `RF` classifier produces reliable classifications through predictions derived from a set of decision trees (Breiman, 2001). Furthermore, this classifier can be successfully used to select and classify variables with the highest discrimination ability between classes. An important factor, considering that the high dimensionality of remote sensing data makes the selection of the most relevant variables a slow process, according to Körting et al. (2013).

The `SVM` classifier is a technique based on statistical learning theory, developed by Vladimir (1995), from studies started in (Vapnik & Chervonenkis, 2015). This learning theory constitutes a series of principles that should be followed in obtaining classifiers, defined as the ability to correctly predict the class of new data from the same domain in which the learning occurred.

14.1.5.7 Cross-validation

The supervised classification was set up with the data being partitioned into 98% for training and 2% for validation of fitted models. The `kfold = 10` method was used in the validation of the results being repeated k-times where each of the k-subsets were used as a test set and other k-1 subsets for training purposes.

The decision about the best algorithm used in target identification can be obtained as a function of model accuracy indicators through cross-validation methodology. The quality of the classification algorithms was also compared in terms of accuracy analysis and kappa metrics (Kohavi, 1995; Kuhn, 2008).

14.1.6 Results and discussion

14.1.6.1 Exploratory analysis

The barplots represented the information of the magnitudes of the spectral targets, according to the classes defined in coffee crops (1), water (2), urban area (3), vegetation (4), exposed soil (5) and pasture (6). Barplots aim to show the relationship between numerical and categorical variables used in the training regions, so it is an efficient way to visually assess the range and variability of the data. In the stacked band barplot, it was possible to compare the training classes according to band magnitudes. For the barplot to Landsat-8, with the exploratory analysis, we observed, a differential effect on the spectral response of each target in the reflective spectrum region. Water had the lowest spectral response followed by forest, coffee crops, grasslands, urban landscape and bare soil (Figure 14.2).

In the case of Sentinel-2, there was a difference compared to Landsat-8, mainly in relation to the red edge and the near-infrared. It is worth noting that the magnitude of the target bands for Sentinel-2 was better distributed compared to Landsat-8, presenting results of similar magnitudes. There was higher energy absorption in the water targets followed by forest, coffee crops, bare soil, urban landscape and grassland landscape (Figure 14.3).

Cordero-Sancho & Sader (2007) examined combinations of spectral bands combined with ancillary data to evaluate the classification accuracy of coffee crops and the nature of spectral confounding

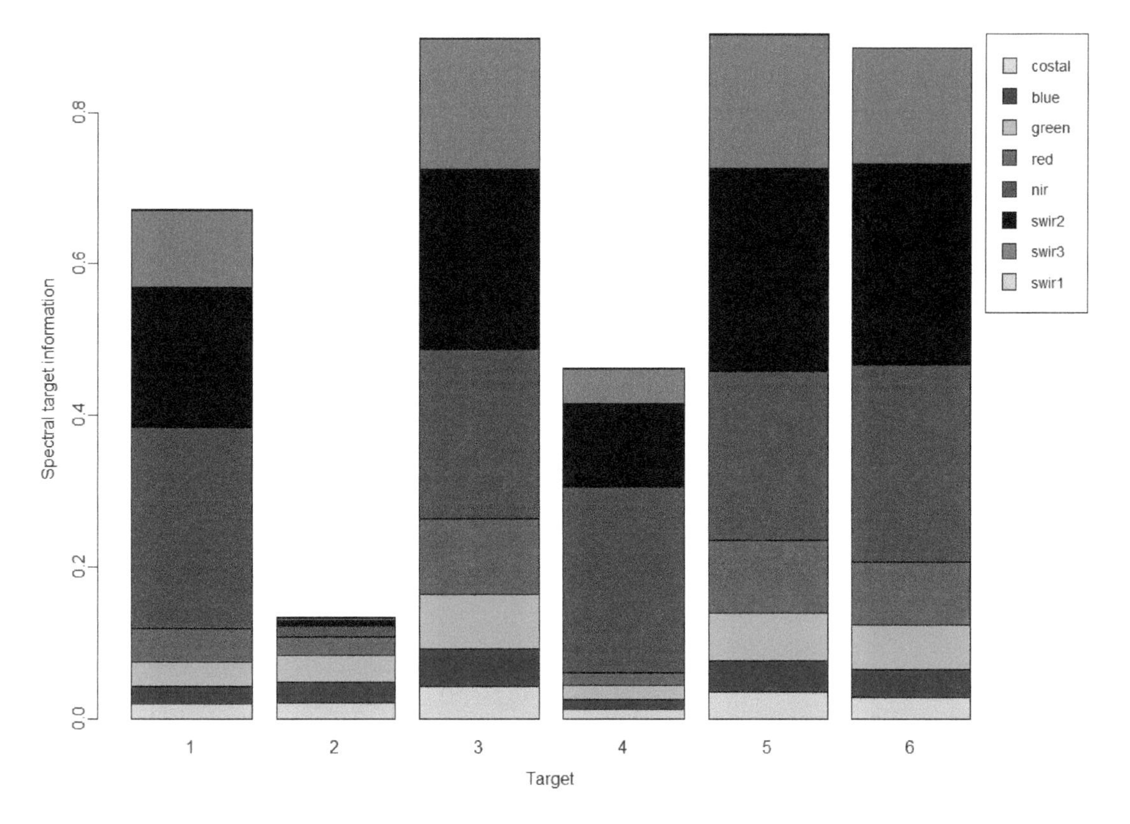

FIGURE 14.2 Using barplots as exploratory analysis of Landsat-8 spectral variables used in supervised classification of targets in coffee crops (1), water (2), urban landscape (3), forest (4), bare soil (5), and grassland (6) in the Funil dam region, state of Minas Gerais, Brazil.

with other woody and non-woody cover types. The supervised classification with only the red, near-infrared, and mid-infrared spectral bands achieved significantly lower classification accuracy compared to other band combinations that included more spectral bands and ancillary data.

The Sentinel-1 backscatter in VH and VV polarizations, calibrated and noise corrected (sigma0), showed most negative values in water targets, followed by bare soil, grassland landscape, coffee crops, urban landscape and forest (Figure 14.4). For each target, there was a difference between the classes, probably in relation to different target textures. In the case of coffee crops, it is important to emphasize the mixing of pathways with the plants in the planting rows. Depending on the period of the phenological cycle of the coffee trees and age of the plants, the soil is even more evident in the area, especially in the post-harvest period, where the process of defoliation of the plants is greater, increasing the likelihood of spectral mixing and spectral divergence between targets of the same class.

Cordero-Sancho & Sader (2007) reported the complexity of the coffee growing landscape, due to small fields divided by tree fences and the presence of pathways between crops, as well as differences in shade intensity, which make it difficult to select training samples of coffee plants. In addition, coffee was spectrally confused with grasslands and low-density forest classes. According to the authors, the causes of spectral confounding can be attributed to the age of the coffee plants. Young coffee crops have more exposed soil compared to mature crops, young coffee conditions also have spectral similarities with other crops.

Regarding the spectral signature of the targets analyzed from the Landsat-8 spectral bands, high water absorption was found in all wavelengths. Water presented a spectral signature consistent

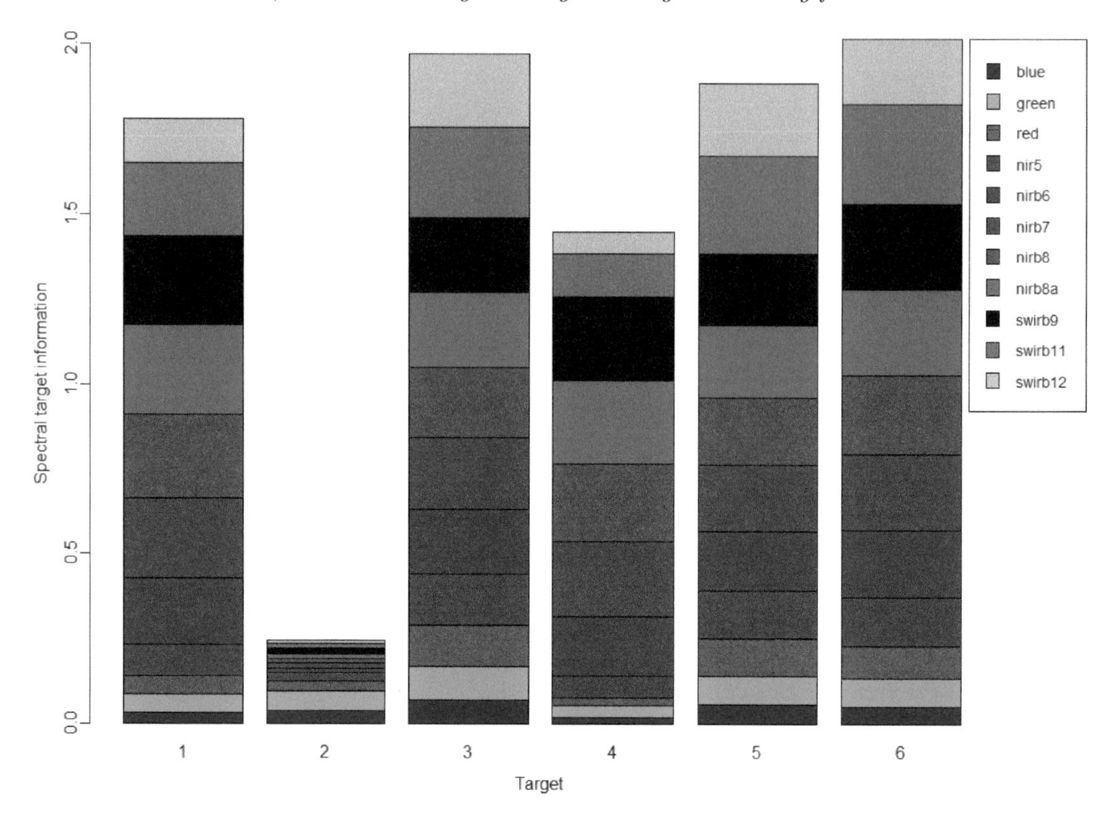

FIGURE 14.3 Using barplots as exploratory analysis of Sentinel-2 spectral variables used in supervised classification of targets in coffee crops (1), water (2), urban landscape (3), forest (4), bare soil (5), and grassland (6) in the Funil dam region, state of Minas Gerais, Brazil.

with reality, absorbing a lot of energy and easily distinguished from other surfaces. There was greater separability of vegetation types in the red and swir2 bands. Targets with forest reflected less than coffee crops and grassland landscape. Bare soil showed a pattern of higher values in the visible and infrared with reduced reflectance only in band 8 (swir3). Urbanized areas reflected more than bare soil in the visible region and less than bare soil in the shortwave infrared (Figure 14.5). In the coffee crops there was higher reflectance compared to the forest, precisely because the soil areas were more apparent than in the vegetation areas, which are denser. The urban area also reflected more than the other classes, which can be explained by the high presence of buildings and constructions in the areas.

Considering the radiant temperature variation of targets, bands 10 and 11 of Landsat-8, bare soil showed the highest temperature, followed by urban landscape, field landscape, coffee crops, water and forest (Figure 14.6). These results were as expected, given the heat islands often observed in urban areas and the milder temperatures observed in water and under forest canopy.

Regarding the spectral signature detected by Sentinel-2 imagery, water showed absorption pattern throughout the spectrum, urban landscape, and soil reflected with progressive increase in wavelength, and the vegetation presented higher absorption in the visible and higher reflectance in the infrared (Figure 14.7). In areas of coffee plantations there was higher reflectance in relation to tree vegetation, as observed in the Landsat-8 imagery.

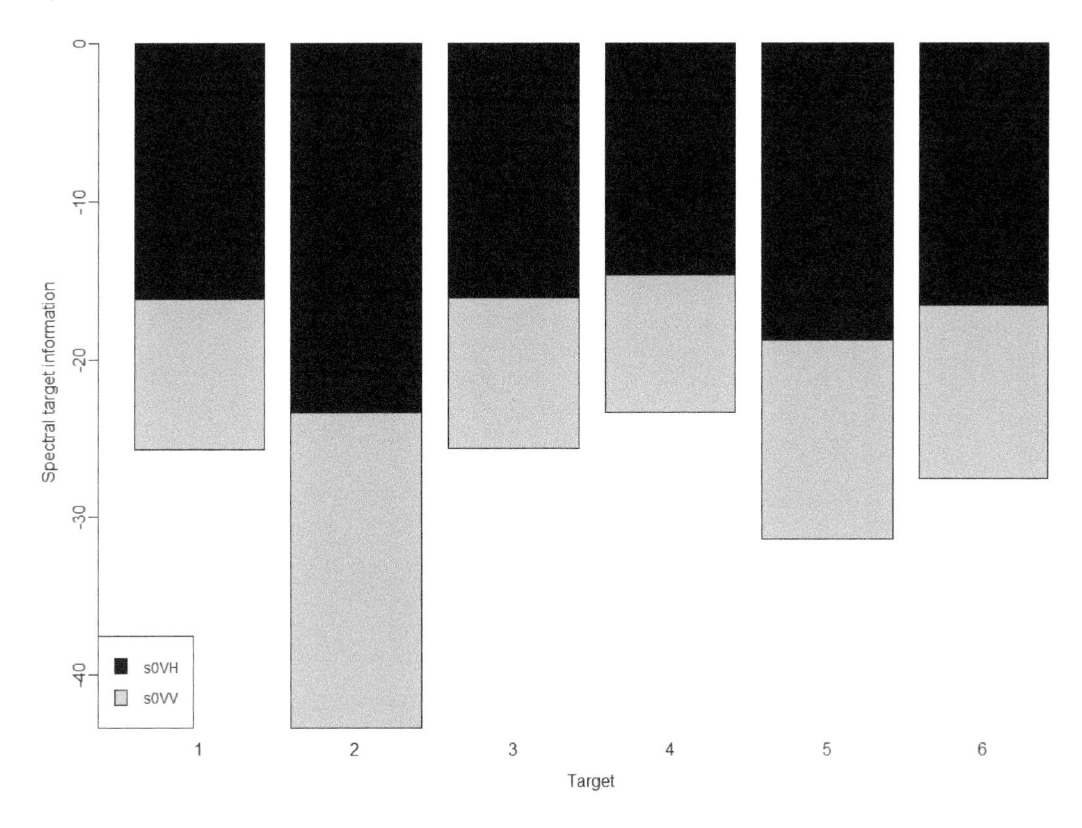

FIGURE 14.4 Using barplots as exploratory analysis of Sentinel-1 spectral variables used in supervised classification of targets in coffee crops (1), water (2), urban landscape (3), forest (4), bare soil (5), and grassland (6) in the Funil dam region, state of Minas Gerais, Brazil.

14.1.6.2 Classification

Based on the results of the supervised classification, we observed that the Sentinel-2 band 8A (red edge) was decisive at the beginning of the classification to separate water from other targets. Afterwards, the NDWI index determined from Landsat-8 data was used to separate coffee crops and forest from urban landscape and bare soil. Subsequently, radiant temperature (band 10, Landsat-8 TIRS), B09 (water vapor), B12 (SWIR) (Sentinel-2), SATVI (Landsat-8 OLI), SAVI (Sentinel-2), and MSAVI (Landsat-8 OLI) were used to separate the remaining targets (Figure 14.8).

Regarding the random forest model, the variables obtained with Landsat-8 MNDWI, tir1, NDWI, GNDVI, SATVI, and the variables obtained with Sentinel-2 imagery, B8A, B12 and B09 were the first 7 variables determining target separability in the Funil dam region (Figure 14.9).

The mapping results were quite similar in terms of spatial pattern of targets observing the whole analyzed area; however, based on visual inspection of the results, the `rpart1SE` algorithm determined prediction of more bare soil and less water when compared to `rf` and `svmLinear2` (Figure 14.10). For a scientific analysis of the results we performed accuracy analysis with the confusion matrix.

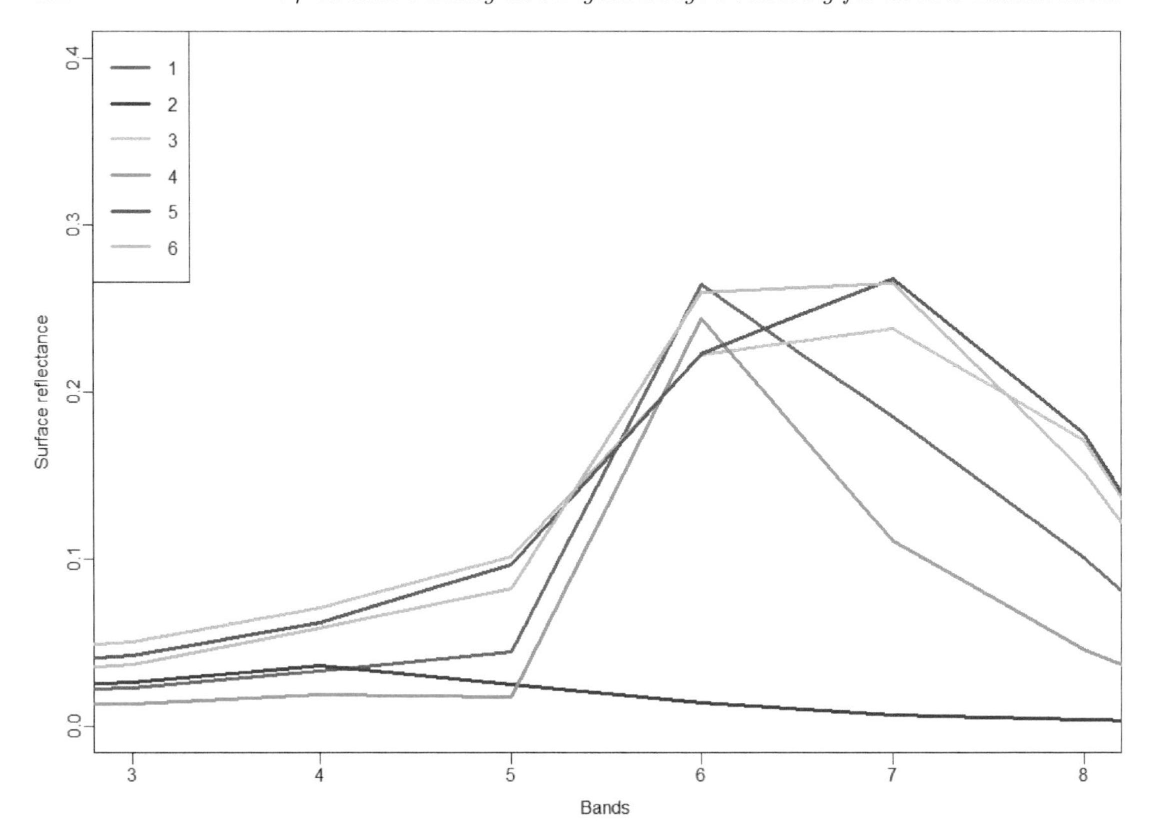

FIGURE 14.5 Spectral signature of blue (3), green (4), red (5), near-infrared (6), shortwave infrared 1 (7) and shortwave infrared 2 (8) bands of Landsat-8 OLI used in supervised classification of targets in coffee crops (1), water (2), urban landscape (3), forest (4), bare soil (5), and grassland (6) in the Funil dam region, state of Minas Gerais, Brazil.

14.1.6.3 Uncertainty Analysis

In the model fitting phase, the highest accuracy and kappa values were found for the `rf` algorithm, followed by `svmLinear2` and `rpart1SE`. However, all the algorithms used showed satisfactory performance to classify the targets, and it is important to evaluate other characteristics of each model to perform the classification for scientific use of thematic results (Figure 14.11).

Although the random forest model was better in the training phase than the support vector machine, in the validation phase the `svmLinear2` algorithm showed better performance based on the statistical results of the confusion matrix (Chapter 13). Therefore, it is considered that for the classification of coffee areas specifically, the `svmLinear2` algorithm was the best indicated for generating thematic mapping of the landscape, in view of the greater number of coffee crops identified and better distinction of vegetation areas as a whole when compared to the other algorithms analyzed. In general, all algorithms used satisfactorily classified the water class. The `rpart1SE` algorithm identified the bare soil areas (5) satisfactorily, compared to the other algorithms. Although random forest had the best accuracy in identifying specific classes, it was noted that `svmLinear2` identified more areas of coffee than the other algorithms. The `rpart1SE` algorithm identified more bare soil, grassland and urban landscape when compared to the other algorithms.

The classification errors between classes of vegetation types evaluated may be related to variations in density and areas of coffee crops, topographic effects, spatial and spectral resolution used in the analysis, different strata of tree height and canopy closure (Cordero-Sancho & Sader, 2007).

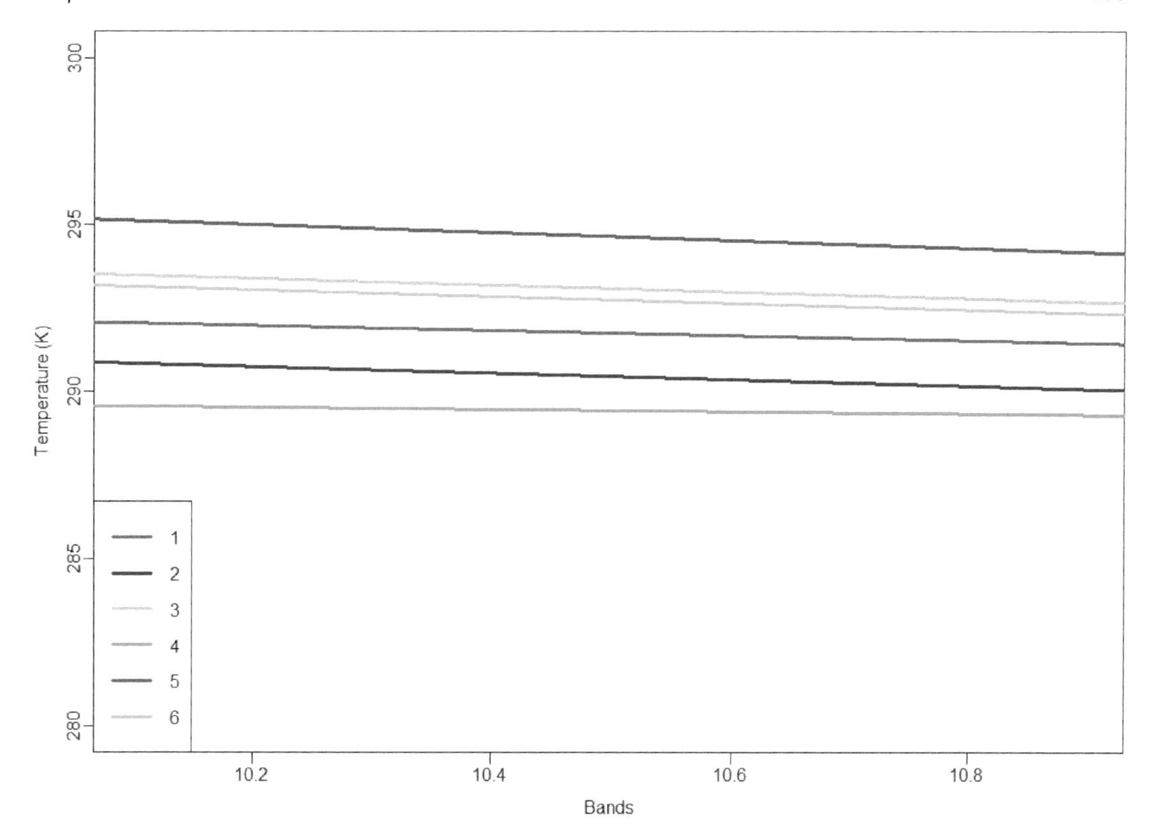

FIGURE 14.6 Spectral signature of thermal infrared spectral bands (10,11) from Landsat-8 TIRS used in supervised classification of targets in coffee crops (1), water (2), urban landscape (3), forest (4), bare soil (5), and grassland (6) in the Funil dam region, state of Minas Gerais, Brazil.

Cordero-Sancho & Sader (2007) reported the difficulties faced during the classification process. According to the authors, for all combinations of bands and post-processing levels applied, the error matrices generally presented the same types of classification errors. Mainly crops and pastures, especially in pastures with sparse presence of trees, were confused with shaded coffee and full sun coffee. Other misclassifications occurred between high-density forest and low-density forest classes, and low-density forest and shaded coffee classes.

One of the notable differences between coffee and other types of woody cover is the canopy height of shading trees and the uniformity in height of coffee plants. Therefore, a more accurate analysis for coffee crop determination may be possible with the use of Light Detection and Ranging (LIDAR) technology, which captures different return signals to differentiate vegetation (Cordero-Sancho & Sader, 2007).

Another alternative to overcome this remote sensing classification problem where coffee and vegetation are mixed is to use local expertise, better known as community mapping. This type of mapping can incorporate geospatial information such as sketch maps, participatory three-dimensional models, aerial photographs, satellite images and tools such as global positioning systems and geographic information systems to compile virtual or physical 2- or 3-dimensional maps (Corbett & Keller, 2006; Rambaldi et al., 2006) and with this improve the classification with hybrid classification methodology (Moreira, 2011).

In the three analyzed classifiers, there were high values of accuracy both in training and validation of the results. Future studies can be conducted to improve the automation of training data in a multisensor approach, enabling a more balanced representation of selected training data for each

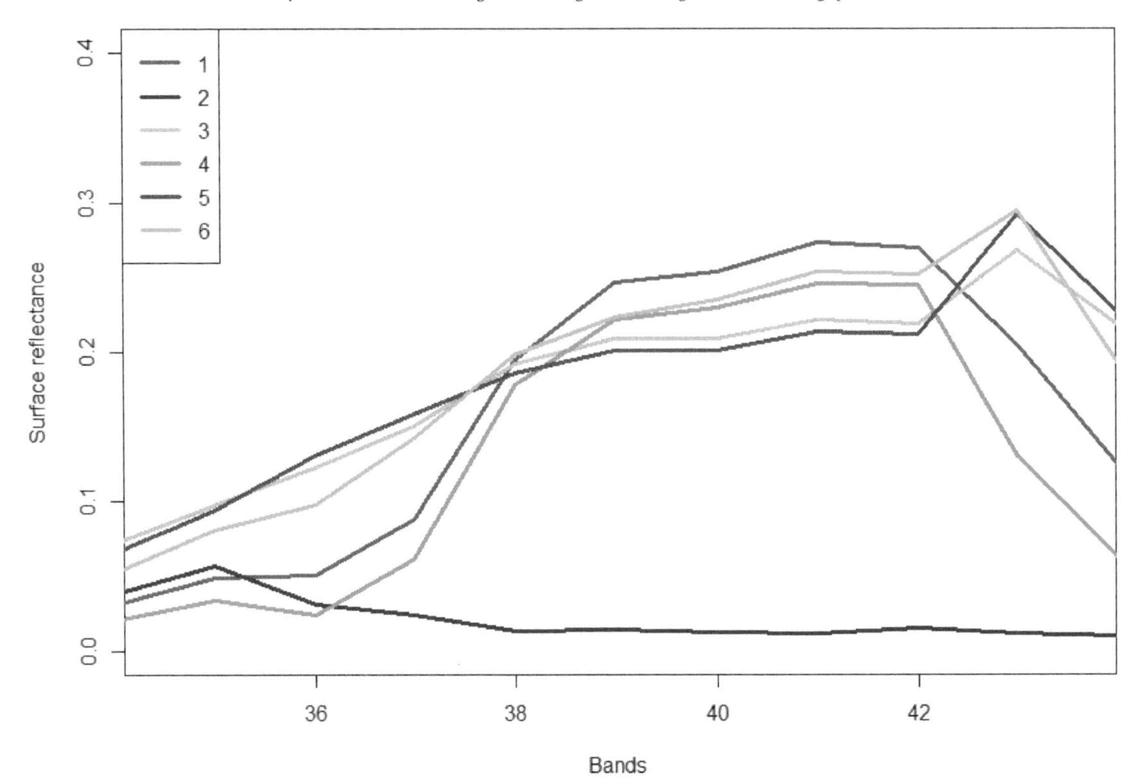

FIGURE 14.7 Spectral signature of blue (35), green (36), red (37), near-infrared-B5 (38), near-infrared-B6 (39), near-infrared-B7 (40), near-infrared-B8 (41), near-infrared-B8A (42), mid-infrared-B11 (43) and Sentinel-2 mid-infrared-B12 (44) used in supervised classification of the targets coffee crops (1), water (2), urban landscape (3), forest (4), bare soil (5), and grassland (6) in the Funil dam region, state of Minas Gerais, Brazil.

target type. It should be considered that the success of any image classification also depends on the analysis of other classification algorithms in order to choose another more suitable classifier procedure (Lu & Weng, 2007).

The random forest classifier can be successfully used to select and classify the variables with the greatest ability to distinguish between targets. This is an important asset, since the high dimensionality of remote sensing data makes the selection of the most relevant variables a time-consuming process (Körting et al., 2013), which is often prone to errors and subjective tasks (Belgiu et al., 2014).

Several studies have systematically investigated the use of the random forest algorithm for hyperspectral data classification (Ham et al., 2005) and Enhanced Thematic Mapper (ETM+) land cover classification (Pal, 2005), or Multispectral Scanner (MSS) and Digital Elevation Model (DEM) data (Gislason et al., 2006). The random forest classifier has been successfully used to map land cover classes (Colditz, 2015; Haas & Ban, 2014; Stefanski et al., 2013; Tsutsumida & Comber, 2015), to map biomass using Landsat temporal data (Frazier et al., 2014), to identify tree health by IKONOS data (Wang et al., 2015), and to map canopy cover and tree biomass using single and multi-temporal Landsat-8 imagery (Karlson et al., 2015).

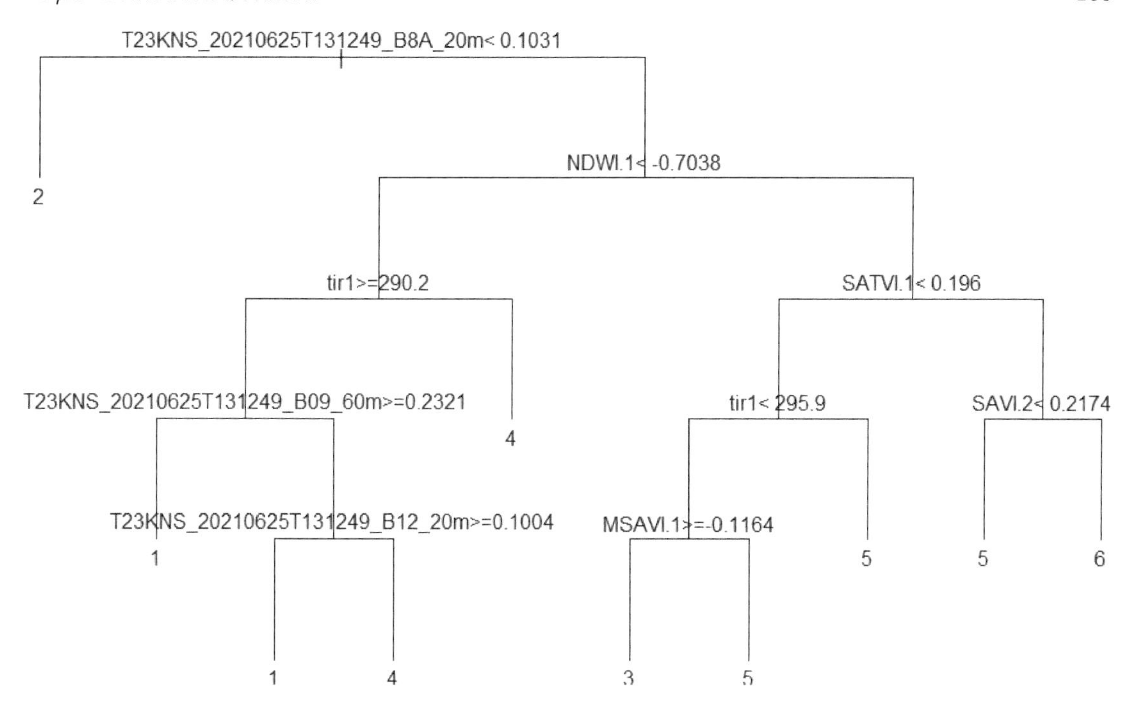

FIGURE 14.8 Decision tree used by the `rpart1SE` model in the supervised classification of the targets coffee crop (1), water (2), urban landscape (3), forest (4), bare soil (5), and grassland landscape (6) in the Funil dam region, state of Minas Gerais, Brazil.

Notably, the higher resolution and image detail in color compositions made by Sentinel-2 compared to Landsat-8 in some bands of the reflective spectrum is remarkable. However, based on the Gini index, the Landsat-8 imaging variables were more important than those of Sentinel-2 in the random forest classification. This can be explained by the dilution of information regarding the data support used to determine a radiance response signal at the time of surface imaging. The 30-m support in the reflective spectrum and 100-m support in the thermal spectrum facilitated pattern recognition by the classifier.

14.1.7 Conclusions

The analyzed classifiers presented satisfactory results to classify landscape targets in the Funil dam region, Minas Gerais, Brazil. Considering the classification of coffee areas, specifically, the `svmLinear2` algorithm can be indicated, in view of the higher number of identified areas and the better distinction of vegetation areas.

The Landsat-8 imagery data were preferred by the `rf` algorithm; however, in the `rpart1SE` algorithm, the Sentinel-2 NIR band 8A (20 m) was fundamental for separating water from the other targets analyzed by establishing a threshold of 0.1031 in the first decision node.

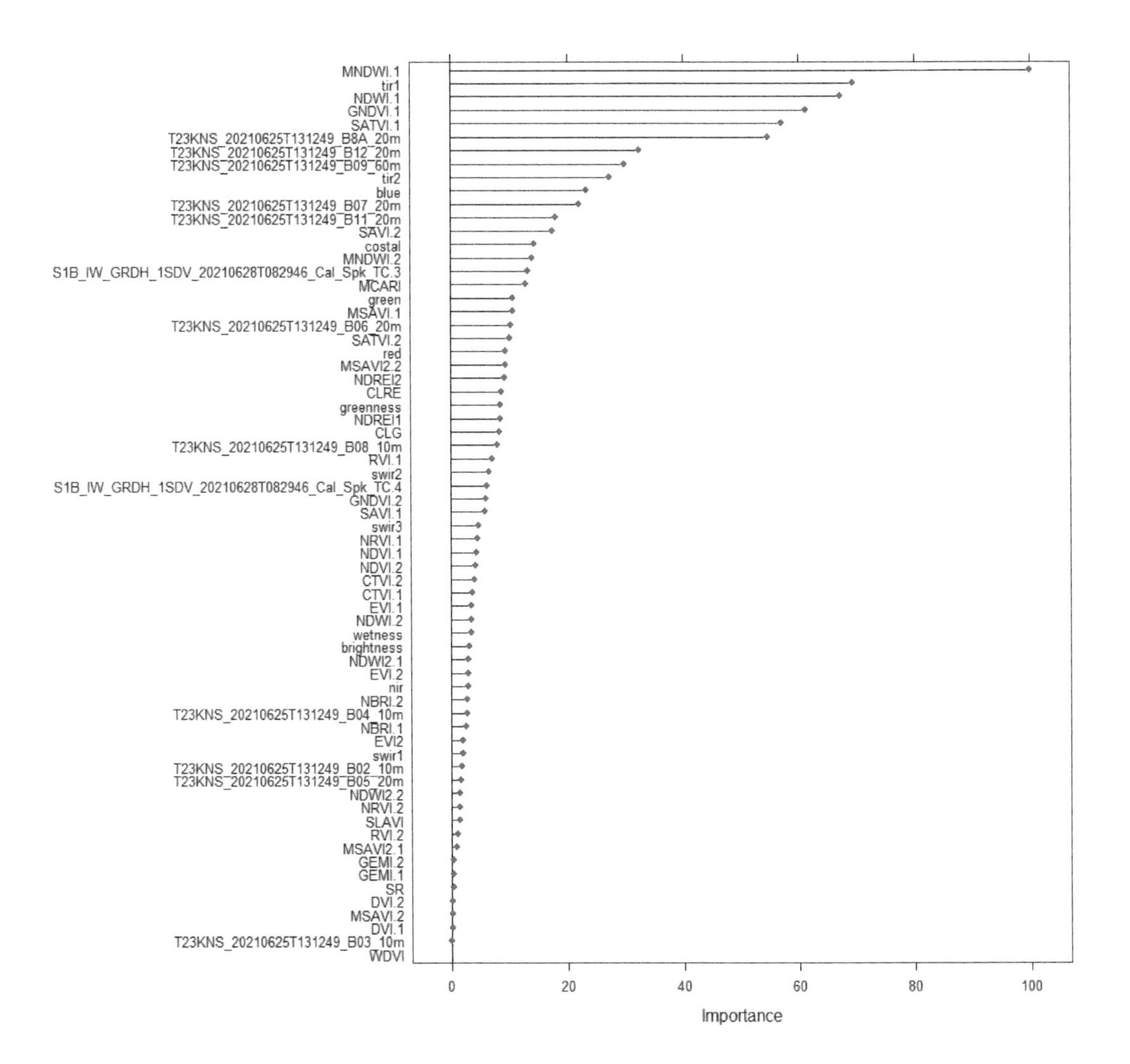

FIGURE 14.9 Importance of variables used by the random forest model in pixel-by-pixel supervised classification of targets in coffee crops (1), water (2), urban landscape (3), forest (4), bare soil (5), grassland (6) in the Funil dam region, state of Minas Gerais, Brazil.

14.2 Solved Exercises

14.2.1 List the topics used in writing a scientific paper in the remote sensing area.

A: Title; Authors; Introduction; Objectives; Methodology; Results; Conclusions; References.

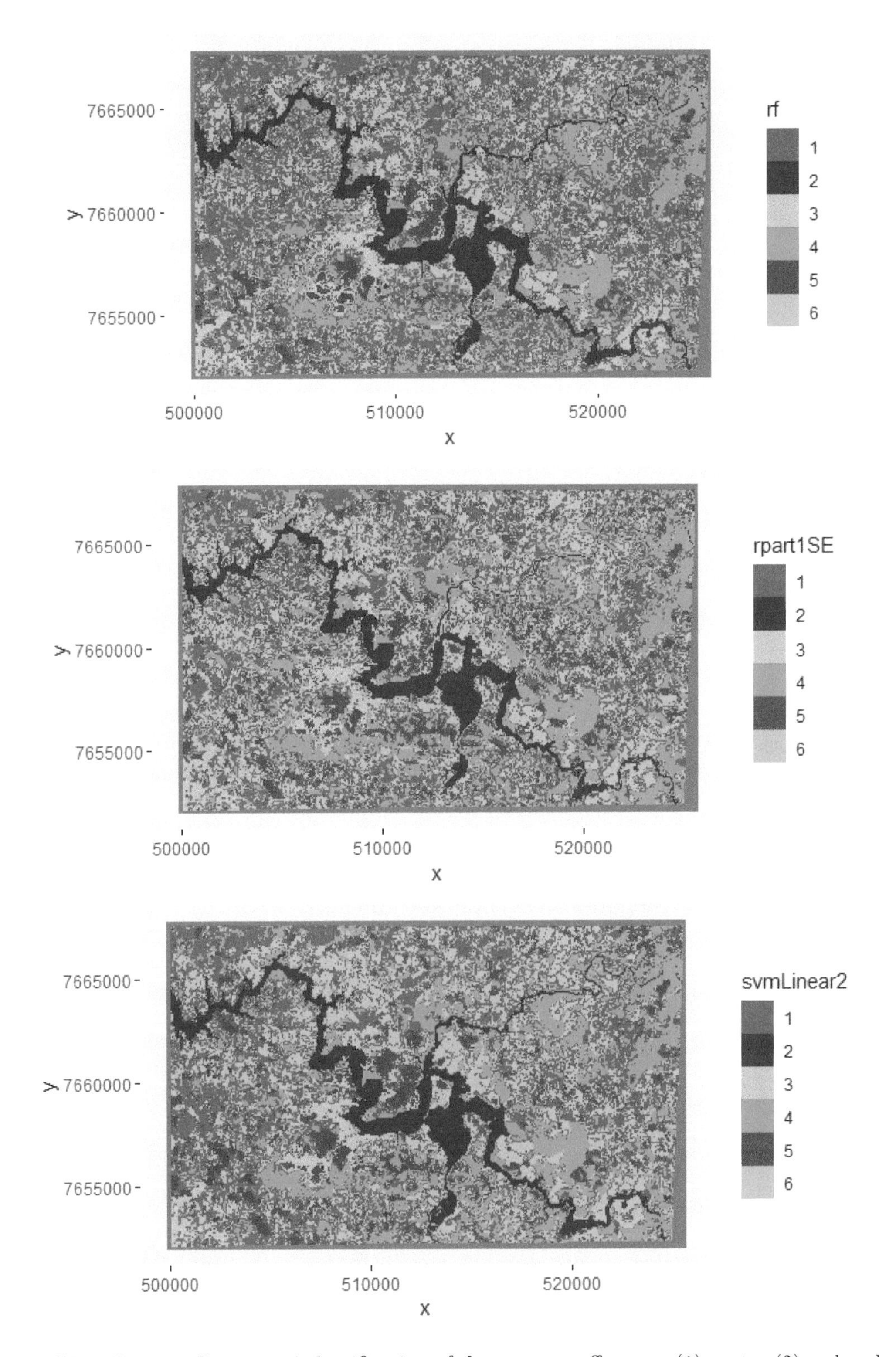

FIGURE 14.10 Supervised classification of the targets coffee crop (1), water (2), urban landscape (3), forest (4), bare soil (5), and grassland landscape (6) with the machine learning algorithms: `rf`, `rpart1SE`, and `svmLinear2`, in the Funil dam region, state of Minas Gerais, Brazil.

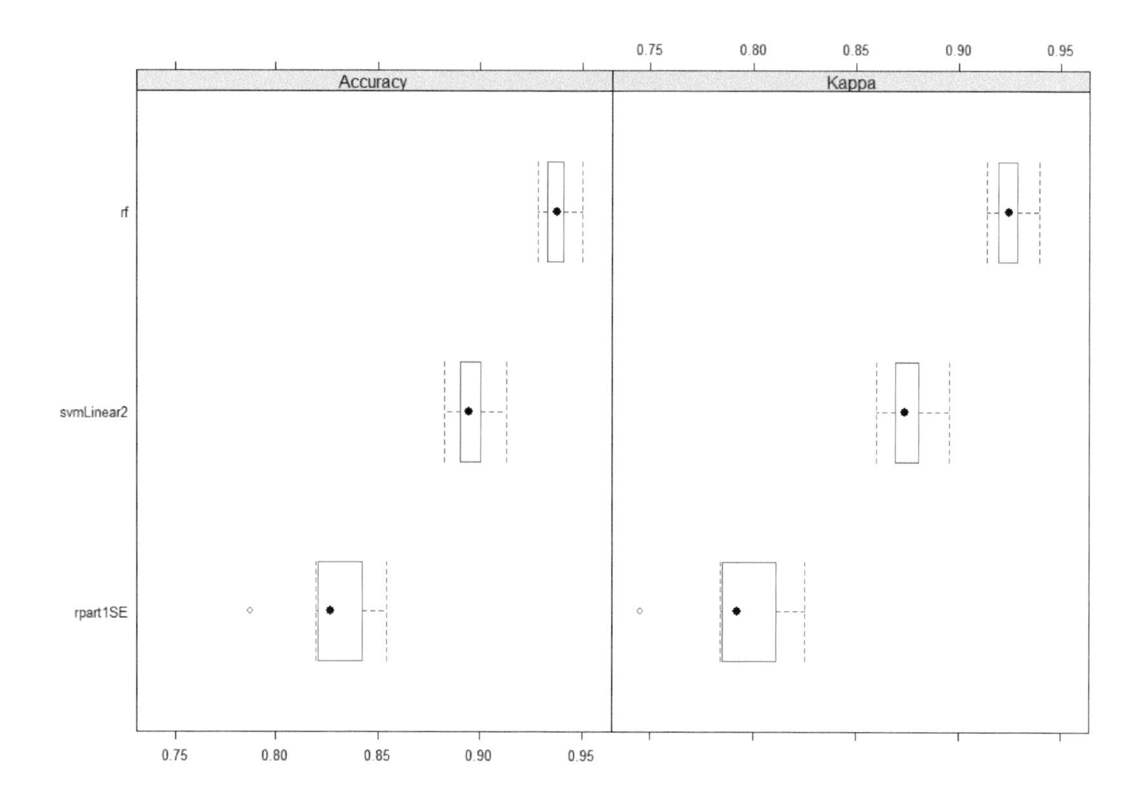

FIGURE 14.11 Accuracy analysis and kappa index in model fitting after supervised classification of remote sensing big data to compare `rf`, `svmLinear2`, and `rpart1SE` algorithms in predicting the targets coffee crops, water, urban landscape, forest, bare soil, and grassland landscape in the Funil dam region, state of Minas Gerais, Brazil.

14.2.2 In a scientific paper, the scientific question to be answered with the research conducted can also be checked in the item of the paper related to:

a. Objective
b. Hypothesis [X]
c. Keywords
d. Methods
e. None of the previous alternatives

14.2.3 How can a scientific paper be written with RStudio?

A: In RStudio there are a number of packages that can create the manuscript according to the submission template of different journals (e.g. `rticles` package). Dependencies of other installed packages enable you to compile the text into different dissemination formats such as LaTeX PDF and to automate cited references in the text with BibTeX and R Markdown (`rmarkdown` package).

References

Abbott, D. (2021). *Top 10 programming languages to learn for 2022.* https://www.linkedin.com/pulse/top-10-programming-languages-learn-2022-daniel-abbott

Achanta, R., Shaji, A., Smith, K., Lucchi, A., Fua, P., & Susstrunk, S. (2011). SLIC Superpixels compared to state-of-the-art superpixel methods. *Journal of Latex Class Files*, *6*(1), 1–16. https://doi.org/10.1109/TPAMI.2012.120

Al-Ahmadi, F. S., & Al-Hames, A. S. (2009). Comparison of four classification methods to extract land use and land cover from raw satellite images for some remote arid areas, Kingdom of Saudi Arabia. *Earth Sciences*, *20*(1).

Ammar, R. (2019). *randomcoloR: Generate Attractive Random Colors* (R package version 1.1.0.1). https://cran.r-project.org/web/packages/randomcoloR/

Appelhans, T., Detsch, F., Woellauer, C. R. S., Forteva, S., Nauss, T., Pebesma, E., Russell, K., Sumner, M., Darley, J., Roudier, P., Schratz, P., Marburg, E. I., & Busetto, L. (2020). *mapview: Interactive Viewing of Spatial Data in R* (R package version 2.11.0.9002). https://github.com/r-spatial/mapview/

Auguie, B., & Antonov, A. (2017). *gridExtra: Miscellaneous Functions for "Grid" Graphics* (R package version 2.3). https://cran.r-project.org/web/packages/gridExtra/

Aybar, C., Qiusheng, W., Bautista, L., Yali, R., Barja, A., Ushey, K., Ooms, J., Appelhans, T., Allaire, J., Tang, Y., Roy, S., Adauto, M., Carrasco, G., Bengtsson, H., Hollister, J., Donchyts, G., & Appel, M. (2022). *rgee: R Bindings for Calling the 'Earth Engine' API* (R package version 1.1.4). https://cran.r-project.org/web/packages/rgee/

Barry, R. G., & Chorley, R. J. (2010). *Atmosphere, weather and climate* (9th ed., p. 533). Routledge.

Becker, R. A., & Chambers, J. M. (1984). *S: An interactive environment for data analysis and graphics* (p. 550). CRC Press.

Belgiu, M., & Csillik, O. (2018). Sentinel-2 cropland mapping using pixel-based and object-based time-weighted dynamic time warping analysis. *Remote Sensing of Environment*, *204*, 509–523. https://doi.org/10.1016/j.rse.2017.10.005

Belgiu, M., Dragut, L., & Strobl, J. (2014). Quantitative evaluation of variations in rule-based classifications of land cover in urban neighbourhoods using WorldView-2 imagery. *ISPRS Journal of Photogrammetry and Remote Sensing*, *87*, 205–215. https://doi.org/10.1016/j.isprsjprs.2013.11.007

Bernardes, T., Moreira, M. A., Adami, M., Giarolla, A., & Rudorff, B. F. T. (2012). Monitoring biennial bearing effect on coffee yield using MODIS remote sensing imagery. *Remote Sensing*, *4*(9), 2492–2509. https://doi.org/10.3390/rs4092492

Bivand, R., Keitt, T., Rowlingson, B., Pebesma, E., Sumner, M., Hijmans, R. J., Baston, D., Rouault, E., Warmerdam, F., Ooms, J., & Rundel, C. (2021). *rgdal: Bindings for the 'Geospatial' Data Abstraction Library* (R packdage version 1.5-19). http://rgdal.r-forge.r-project.org

Blaschke, T. (2010). Object based image analysis for remote sensing. *ISPRS Journal of Photogrammetry and Remote Sensing*, *65*(1), 2–16. https://doi.org/10.1016/j.isprsjprs.2009.06.004

Bolanos, S. (2007). *Using image analysis and GIS for coffee mapping* (p. 131) [PhD thesis]. McGill University.

Bourgoin, C., Oszwald, J., Bourgoin, J., Gond, V., Blanc, L., Dessard, H., Phan, T. V., Sist, P., Läderach, P., & Reymondin, L. (2020). Assessing the ecological vulnerability of forest landscape to agricultural frontier expansion in the Central Highlands of Vietnam. *International Journal of Applied Earth Observation and Geoinformation*, *84*, 101958. https://doi.org/10.1016/j.jag.2019.101958

Breiman, L. (2001). Random forests. *Machine Learning, 45*(1), 5–32. https://doi.org/10.1023/A:1010933404324

Büttner, G. (2014). CORINE land cover and land cover change products. In I. Manakos & M. Braun (Eds.), *Land use and land cover mapping in Europe* (pp. 55–74). Springer.

Campbell, J. B., & Wynne, R. H. (2011). *Introduction to remote sensing* (5th ed., p. 667). The Guilford Press.

Claverie, M., Ju, J., Masek, J. G., Dungan, J. L., Vermote, E. F., Roger, J.-C., Skakun, S. V., & Justice, C. (2018). The harmonized Landsat and Sentinel-2 surface reflectance data set. *Remote Sensing of Environment, 219*, 145–161. https://doi.org/10.1016/j.rse.2018.09.002

Colditz, R. (2015). An evaluation of different training sample allocation schemes for discrete and continuous land cover classification using decision tree-based algorithms. *Remote Sensing, 7*(8), 9655–9681. https://doi.org/10.3390/rs70809655

Corbett, J., & Keller, P. (2006). Using community information systems to communicate traditional knowledge embedded in the landscape. *Participatory Learning and Action, 54*(1), 21–27.

Cordero-Sancho, S., & Sader, S. A. (2007). Spectral analysis and classification accuracy of coffee crops using Landsat and a topographic-environmental model. *International Journal of Remote Sensing, 28*(7), 1577–1593. https://doi.org/10.1080/01431160600887680

Csillik, O. (2017). Fast segmentation and classification of very high resolution remote sensing data using SLIC superpixels. *Remote Sensing, 9*(3), 1–19. https://doi.org/10.3390/rs9030243

Dalgaard, P. (2008). *Introductory Statistics with R* (2nd ed., p. 364). Springer.

Dantas, A. A. A., Carvalho, L. G., & Ferreira, E. (2007). Climatic classification and tendencies in Lavras region, MG. *Ciência e Agrotecnologia, 31*(6), 1862–1866. https://doi.org/10.1590/S1413-70542007000600039

Douglas, A., Roos, D., Mancini, F., Couto, A., & Lusseau, D. (2022). *An introduction to R* (pp. 27–52). https://doi.org/10.1201/b15352-7

Drusch, M., Del Bello, U., Carlier, S., Colin, O., Fernandez, V., Gascon, F., Hoersch, B., Isola, C., Laberinti, P., Martimort, P., Meygret, A., Spoto, F., Sy, O., Marchese, F., & Bargellini, P. (2012). Sentinel-2: ESA's optical high-resolution mission for GMES operational services. *Remote Sensing of Environment, 120*, 25–36. https://doi.org/10.1016/j.rse.2011.11.026

Einav, L., & Levin, J. (2014). Economics in the age of big data. *Science, 346*, 6210. https://doi.org/10.1126/science.1243089

Esch, T., Thiel, M., Bock, M., Roth, A., & Dech, S. (2008). Improvement of image segmentation accuracy based on multiscale optimization procedure. *IEEE Geoscience and Remote Sensing Letters, 5*(3), 463–467. https://doi.org/10.1109/LGRS.2008.919622

Farr, T. G., Rosen, P. A., Caro, E., Crippen, R., Duren, R., Hensley, S., Kobrick, M., Paller, M., Rodriguez, E., Roth, L., Seal, D., Shaffer, S., Shimada, J., Umland, J., Werner, M., Oskin, M., Burbank, D., & Alsdorf, D. (2007). The Shuttle Radar Topography Mission. *Reviews of Geophysics, 45*(2), RG2004. https://doi.org/10.1029/2005RG000183

Fox, J. (2009). Aspects of the social organization and trajectory of the R project. *R J., 1*(2), 5.

Frazier, R. J., Coops, N. C., Wulder, M. A., & Kennedy, R. (2014). Characterization of aboveground biomass in an unmanaged boreal forest using Landsat temporal segmentation metrics. *ISPRS Journal of Photogrammetry and Remote Sensing, 92*, 137–146. https://doi.org/10.1016/j.isprsjprs.2014.03.003

Funk, C., Peterson, P., Landsfeld, M., Pedreros, D., Verdin, J., Shukla, S., Husak, G., Rowland, J., Harrison, L., & Hoell, A. (2015). The climate hazards infrared precipitation with stations—a new environmental record for monitoring extremes. *Scientific Data, 2*(1), 1–21. https://doi.org/10.1038/sdata.2015.66

Gallego, F. J., Kussul, N., Skakun, S., Kravchenko, O., Shelestov, A., & Kussul, O. (2014). Efficiency assessment of using satellite data for crop area estimation in Ukraine. *International Journal of Applied Earth Observation and Geoinformation, 29*, 22–30. https://doi.org/10.1016/j.jag.2013.12.013

Ghosh, A., Mandel, A., Kenduiywo, B., & Hijmans, R. J. (2021). *luna: Tools for satellite remote sensing (Earth Observation) data processing* (R package version 0.3-3). https://rdrr.io/github/rspatial/luna/

Gislason, P. O., Benediktsson, J. A., & Sveinsson, J. R. (2006). Random forests for land cover classification. *Pattern Recognition Letters*, *27*(4), 294–300. https://doi.org/10.1016/j.patrec.2005.08.011

Goes, J. C., & Oliveira, E. S. (2022). Relationship between soil management and rainfall erosion: A case study in the Guajará-Mirim River watershed, Vigia-Pará. *International Journal of Advanced Engineering Research and Science*, *9*, 5. https://doi.org/10.22161/ijaers.95.16

Greenberg, J. A. (2000). *Gdalutils* (R package version 2.0.3.2). https://github.com/cran/gdalUtils

Haas, J., & Ban, Y. (2014). Urban growth and environmental impacts in Jing-Jin-Ji, the Yangtze, River Delta and the Pearl River Delta. *International Journal of Applied Earth Observation and Geoinformation*, *30*, 42–55. https://doi.org/10.1016/j.jag.2013.12.012

Ham, J., Yangchi Chen, Crawford, M. M., & Ghosh, J. (2005). Investigation of the random forest framework for classification of hyperspectral data. *IEEE Transactions on Geoscience and Remote Sensing*, *43*(3), 492–501. https://doi.org/10.1109/TGRS.2004.842481

Hanson, B. A. (2017). *SpecHelpers: Spectroscopy Related Utilities* (R package version 0.2.7). https://cran.r-project.org/web/packages/SpecHelpers/

Higa, L., Marcato Junior, J., Rodrigues, T., Zamboni, P., Silva, R., Almeida, L., Liesenberg, V., Roque, F., Libonati, R., Gonçalves, W. N., & Silva, J. A. (2022). Active fire mapping on Brazilian Pantanal based on deep learning and CBERS 04A imagery. *Remote Sensing*, *14*(3), 688. https://doi.org/10.3390/rs14030688

Hijmans, R. J., Bivand, R., Pebesma, E., & Sumner, M. D. (2022). *terra: Spatial Data Analysis* (R package version 1.6-7). https://rspatial.org/terra/

Hijmans, R. J., Etten, J. van, Sumner, M., Cheng, J., Baston, D., Bevan, A., Bivand, R., Busetto, L., Canty, M., Fasoli, B., Forrest, D., Ghosh, A., Goliche, D., Gray, J., Greenberg, J. A., Hiemstra, P., Hingee, K., Geosciences, I. for M. A., Karney, C., … Wueest, R. (2020). *raster: Geographic Data Analysis and Modeling* (R package version 3.4.5). https://rspatial.org/raster

Hornik, K. (2012a). Are there too many R packages? *Austrian Journal of Statistics*, *41*(1), 59–66. https://doi.org/10.17713/ajs.v41i1.188

Hornik, K. (2012b). The comprehensive R archive network. *Wiley Interdisciplinary Reviews: Computational Statistics*, *4*(4), 394–398. https://doi.org/10.1002/wics.1212

Huete, A. R., Justice, C., & Van Leeuwen, W. (1999). *MODIS vegetation index (MOD 13): Algorithm theoretical basis document* (Version 3, p. 129). National Aeronautics; Space Administration.

Hunt, D. A., Tabor, K., Hewson, J. H., Wood, M. A., Reymondin, L., Koenig, K., Schmitt-Harsh, M., & Follett, F. (2020). Review of remote sensing methods to map coffee production systems. *Remote Sensing*, *12*(12), 2041. https://doi.org/10.3390/rs12122041

Hussain, S. (2015). Educational data mining using R programming and R Studio. *Journal of Applied and Fundamental Sciences*, *1*(1), 45.

Ihaka, R. (1998). *R : Past and future history* (p. 35). Statistics Department, The University of Auckland.

Ihaka, R., & Gentleman, R. (1996). R: A language for data analysis and graphics. *Journal of Computational and Graphical Statistics*, *5*(3), 299–314. https://doi.org/10.1080/10618600.1996.10474713

Jensen, J. R. (2005). *Introductory digital image processing: A remote sensing perspective* (3rd ed., p. 526). Pearson Prentice Hall.

Karlson, M., Ostwald, M., Reese, H., Sanou, J., Tankoano, B., & Mattsson, E. (2015). Mapping tree canopy cover and aboveground biomass in Sudano-Sahelian woodlands using Landsat 8 and random forest. *Remote Sensing*, *7*(8), 10017–10041. https://doi.org/10.3390/rs70810017

Kawakubo, F. S., & Pérez Machado, R. P. (2016). Mapping coffee crops in southeastern Brazil using spectral mixture analysis and data mining classification. *International Journal of Remote Sensing*, *37*(14), 3414–3436. https://doi.org/10.1080/01431161.2016.1201226

Kintisch, E. (2007). Carbon emissions: Improved monitoring of rainforests helps pierce haze of deforestation. *Science*, *316*(5824), 536–537. https://doi.org/10.1126/science.316.5824.536

Kohavi, R. (1995). A study of cross-validation and bootstrap for accuracy estimation and model selection. *IJCAI'95: Proceedings of the 14th International Joint Conference on Artificial Intelligence*, *2*, 1137–1145.

Köhler, V. B. (1998). Remote sensing in vegetation study. *Boletim de Geografia*, *16*(1), 107–118.

Körting, T. S., Garcia Fonseca, L. M., & Câmara, G. (2013). GeoDMA-Geographic data mining analyst. *Computers & Geosciences*, *57*, 133–145. https://doi.org/10.1016/j.cageo.2013.02.007

Krassenburg, M. (2016). Sentinel-1 mission status. *Proceedings of the 11th European Conference on Synthetic Aperture Radar (EUSAR 2016)*, 1–6.

Kuhn, M. (2008). Building predictive models in R using the caret package. *Journal of Statistical Software*, *28*(5). https://doi.org/10.18637/jss.v028.i05

Kuhn, M., Wing, J., Weston, S., Williams, A., Keefer, C., Engelhardt, A., Cooper, T., Mayer, Z., Kenkel, B., Team, R. C., Benesty, M., Lescarbeau, R., Ziem, A., Scrucca, L., Tang, Y., Candan, C., & Hunt, T. (2020). *caret: Classification and Regression Training* (R package version 6.0-86). https://github.com/topepo/caret/

Kushalappa, A. C., & Eskes, A. B. (1989). *Coffee rust: Epidemiology, resistance, and management* (1st ed., p. 358). CRC Press.

Langford, M., & Bell, W. (1997). Land cover mapping in a tropical hillsides environment: A case study in the Cauca region of Colombia. *International Journal of Remote Sensing*, *18*(6), 1289–1306. https://doi.org/10.1080/014311697218421

Larrañaga, A., & Álvarez-Mozos, J. (2016). On the added value of quad-pol data in a multi-temporal crop classification framework based on RADARSAT-2 imagery. *Remote Sensing*, *8*(4), 335. https://doi.org/10.3390/rs8040335

Leutner, B., Horning, N., Schwalb-Willmann, J., & Hijmans, R. J. (2019). *RStoolbox: Tools for Remote Sensing Data Analysis* (R package version 0.2.6). https://cran.r-project.org/web/packages/RStoolbox/

Lobser, S. E., & Cohen, W. B. (2007). MODIS tasselled cap: Land cover characteristics expressed through transformed MODIS data. *International Journal of Remote Sensing*, *28*(22), 5079–5101. https://doi.org/10.1080/01431160701253303

Lovelace, R., Nowosad, J., & Muenchow, J. (2019). *Geocomputation with R* (1st ed., p. 339). CRC Press.

Lu, D., Batistella, M., Moran, E., & Miranda, E. E. (2008). A comparative study of Landsat TM and SPOT HRG images for vegetation classification in the Brazilian Amazon. *Photogrammetric Engineering & Remote Sensing*, *74*(3), 311–321. https://doi.org/10.14358/PERS.74.3.311

Lu, D., & Weng, Q. (2007). A survey of image classification methods and techniques for improving classification performance. *International Journal of Remote Sensing*, *28*(5), 823–870. https://doi.org/10.1080/01431160600746456

Malenovský, Z., Rott, H., Cihlar, J., Schaepman, M. E., García-Santos, G., Fernandes, R., & Berger, M. (2012). Sentinels for science: Potential of Sentinel-1, -2, and -3 missions for scientific observations of ocean, cryosphere, and land. *Remote Sensing of Environment*, *120*, 91–101. https://doi.org/10.1016/j.rse.2011.09.026

Marinho, J. L. M., Cid, Y. P. M., Barros, W. V. R., Costa, M. S. S., Santos Ribeiro, E., Silva Carneiro, F., Repolho, S. M., Jesus Cordeiro, D. F., Jesus, R. C. S., & Amaral, A. P. M. (2022). Application of geotechnology to identify vegetation fragments in the municipality of Belém-Pará-Brasil in the year 2020. *Research, Society and Development*, *11*(4), e46211426745–e46211426745. https://doi.org/10.33448/rsd-v11i4.26745

Marques, T. A. (2019). *A hands-on tutorial on R and R Studio* (p. 23).

Matias, J. M. J. E. (2019). *Análise comparada das potencialidades e limitações dos dados Sentinel-2 e Landsat-8 para aplicações operacionais em ambiente e planeamento territorial. Caso de estudo: Os municípios de Catumbela e Lobito - Angola* (p. 157) [PhD thesis]. Universidade Nova de Lisboa.

Matter, U. (2021). *A brief introduction to advanced programming with R* (pp. 1–15).

Messias, C. G., & Ferreira, M. M. (2014). Geomorphological study of Funil reservoir watersheds, upper Grande river basin (MG), using stereoscopic images of ALOS/PRISM sensor. *Revista Do Departamento de Geografia*, *28*, 237–262. https://doi.org/10.11606/rdg.v28i0.554

Meyer, D., Dimitriadou, E., Hornik, K., Weingessel, A., Leisch, F., Chang, C.-C., & Lin, C.-C. (2021). *e1071: Misc Functions of the Department of Statistics, Probability Theory Group (Formerly: E1071), TU Wien* (R package version 1.7-6). https://cran.r-project.org/web/packages/e1071/

Moreira, M. A. (2011). *Fundamentos do sensoriamento remoto e metodologias de aplicação* (4th ed., p. 422). Editora UFV.

Muenchen, R. A. (2022). *The popularity of data analysis software*. http://r4stats.com/popularity

Nowosad, J. (2021). Motif: An open-source R tool for pattern-based spatial analysis. *Landscape Ecology*, *36*(1), 29–43. https://doi.org/10.1007/s10980-020-01135-0

Nowosad, J., Mettes, P., & Jekel, C. (2022). *supercells: Superpixels of Spatial Data* (R package version 0.9.1). https://cran.r-project.org/web/packages/supercells/

Nowosad, J., & Stepinski, T. (2021). *Generalizing the Simple Linear Iterative Clustering (SLIC) superpixels* (pp. 1–6). https://doi.org/10.25436/E2QP4R

Oliveira Filho, A. T., Vilela, E. A., Gavilanes, M. L., & Carvalho, D. A. (1994). Comparison of the woody flora and soils of six areas of montane semideciduous forest in Southern Minas Gerais, Brazil. *Edinburgh Journal of Botany*, *51*, 355–389.

Pal, M. (2005). Random forest classifier for remote sensing classification. *International Journal of Remote Sensing*, *26*(1), 217–222. https://doi.org/10.1080/01431160412331269698

Patil, S. (2016). Big data analytics using IoT. *International Research Journal of Engineering and Technology*, *3*(7), 78–81. https://doi.org/10.3390/mol2net-07-09183

Pebesma, E., Bivand, R., Racine, E., Sumner, M., Cook, I., Keitt, T., Lovelac, R., Wickham, H., Ooms, J., Müller, K., Pedersen, T. L., & Baston, D. (2021). *sf: Simple Features for R* (R package version 0.9-7). https://cran.r-project.org/web/packages/sf/

Pebesma, E., Sumner, M., Racine, E., Fantini, A., & Blodgett, D. (2022). *stars: Spatiotemporal Arrays, Raster and Vector Data Cubes* (R package version 0.6-0). https://cran.r-project.org/web/packages/stars/

Pereira, R. H. M., Goncalves, C. N., Araujo, P. H. F., Carvalho, G. D., Arruda, R. A., Nascimento, I., Costa, B. S. P., Cavedo, W. S., Andrade, P. R., Silva, A., Braga, C. K. V., Schmertmann, C., Samuel-Rosa, A., & Ferreira, D. (2021). *geobr: Download Official Spatial Data Sets of Brazil*. https://github.com/cran/geobr

Prodes Amazônia. (2020). *PRODES Amazônia: Monitoramento do desmatamento da floresta Amazônica Brasileira por satélite*. http://www.obt.inpe.br/OBT/assuntos/programas/amazonia/prodes

Qadir, A., & Mondal, P. (2020). Synergistic use of radar and optical satellite data for improved monsoon cropland mapping in India. *Remote Sensing*, *12*(3), 522. https://doi.org/10.3390/rs12030522

R Core Team. (2020). *R: A language and environment for statistical computing*. R Foundation for Statistical Computing. https://www.R-project.org/

Rambaldi, G., Chambers, R., McCall, M., & Fox, J. (2006). Practical ethics for PGIS practitioners, facilitators, technology intermediaries and researchers. *Participatory Learning and Action*, *54*(1), 106–113.

Reudenbach, C., Meyer, H., Detsch, F., Möller, F., Nauss, T., Opgenoorth, L., & Marburg, E. I. (2022). *uavRst: Unmanned Aerial Vehicle R Tools* (R package version 0.5-4). https://github.com/gisma/uavRst/

Rodríguez, G. (2022). Introducing R. *R Programming for Bioinformatics*, 13–16. https://doi.org/10.1201/9781420063684-4

Rudorff, B. F. T., Shimabukuro, Y. E., & Ceballos, J. C. (2007). *O sensor MODIS e suas aplicações ambientais no Brasil* (p. 423). Editora Parêntese.

Sanatan, V. (2017). Tutorial R e RStudio. *Universidade de Sao Paulo*, *11*, 33.

Santos, H. G., Jacomine, P. K. T., Anjos, L. H. C., Oliveira, V. A., Lumbreras, J. F., Coelho, M. R., Almeida, J. A., Araujo Filho, J. C., Oliveira, J. B., & Cunha, T. J. F. (2018). *Sistema Brasileiro de Classificação de Solos* (5th ed., p. 355). Empresa Brasileira de Pesquisa Agropecuária; Embrapa-Solos.

Sarkarand, D., Andrews, F., Wright, K., Klepeis, N., Larsson, J., & Murrell, P. (2021). *lattice: Trellis Graphics for R* (R package version 0.20-45). https://cran.r-project.org/web/packages/lattice/

Sarvary, M. A. (2014). Biostatistics in the classroom: Teaching introductory biology students how to use the statistical software 'R' effectively. *Proceedings of the Association for Biology Laboratory Education, 35*, 405–407.

Schmitt-Harsh, M. (2013). Landscape change in Guatemala: Driving forces of forest and coffee agroforest expansion and contraction from 1990 to 2010. *Applied Geography, 40*, 40–50. https://doi.org/10.1016/j.apgeog.2013.01.007

Seyednasrollah, B., Kumar, M., & Link, T. E. (2013). On the role of vegetation density on net snow cover radiation at the forest floor. *Journal of Geophysical Research: Atmospheres, 118*(15), 8359–8374. https://doi.org/10.1002/jgrd.50575

Seyednasrollah, B., Milliman, T., & Richardson, A. D. (2021). *xROI: Delineate Region of Interests (ROI's) and Extract Time-Series Data from Digital Repeat Photography Images* (R package version 0.9.20). https://cran.r-project.org/web/packages/xROI/

Sousa, K., Sparks, A. H., Ghosh, A., Peterson, P., Ashmall, W., Etten, J. van, & Solberg, S. Ø. (2022). *chirps: API Client for CHIRPS and CHIRTS* (R package version 0.1.4). https://cran.r-project.org/web/packages/chirps/

Stefanski, J., Mack, B., & Waske, O. (2013). Optimization of object-based image analysis with random forests for land cover mapping. *IEEE Journal of Selected Topics in Applied Earth Observations and Remote Sensing, 6*(6), 2492–2504. https://doi.org/10.1109/JSTARS.2013.2253089

Stutz, D., Hermans, A., & Leibe, B. (2018). Superpixels: An evaluation of the state-of-the-art. *Computer Vision and Image Understanding, 166*(October), 1–27. https://doi.org/10.1016/j.cviu.2017.03.007

Taylor, C. C. (2018). Using R to teach statistics. *Proceedings of the Tenth International Conference on Teaching Statistics - Icots10.*

Tennekes, M., Nowosad, J., Gombin, J., Jeworutzki, S., Russell, K., Zijdeman, R., Clouse, J., Lovelace, R., & Muenchow, J. (2020). *tmap: Thematic Maps* (R package version 3.2). https://cran.r-project.org/web/packages/tmap/

Teucher, A., Russell, K., & Bloch, M. (2022). *rmapshaper: Client for 'mapshaper' for 'Geospatial' Operations* (R package version 0.4.6). https://cran.r-project.org/web/packages/rmapshaper/

The R Foundation. (2022). *CRAN Mirrors.* https://cran.r-project.org/mirrors

Thenkabail, P. S., & Wu, Z. (2012). An Automated Cropland Classification Algorithm (ACCA) for Tajikistan by combining Landsat, MODIS, and secondary data. *Remote Sensing, 4*(10), 2890–2918. https://doi.org/10.3390/rs4102890

Tian, S., Zhang, X., Tian, J., & Sun, Q. (2016). Random forest classification of eetland landcovers from multi-sensor data in the arid region of Xinjiang, China. *Remote Sensing, 8*(11), 954. https://doi.org/10.3390/rs8110954

Tsutsumida, N., & Comber, A. J. (2015). Measures of spatio-temporal accuracy for time series land cover data. *International Journal of Applied Earth Observation and Geoinformation, 41*, 46–55. https://doi.org/10.1016/j.jag.2015.04.018

USGS. (2021). *Earth Explorer.* https://earthexplorer.usgs.gov

Vapnik, V. N., & Chervonenkis, A. Ya. (2015). On the uniform convergence of relative frequencies of events to their probabilities. In *Measures of complexity* (pp. 11–30). Springer International Publishing. https://doi.org/10.1007/978-3-319-21852-6

Vladimir, N. V. (1995). *The nature of statistical learning theory.* Springer.

Wang, H., Zhao, Y., Pu, R., & Zhang, Z. (2015). Mapping Robinia Pseudoacacia Forest Health Conditions by Using Combined Spectral, Spatial, and Textural Information Extracted from IKONOS Imagery and Random Forest Classifier. *Remote Sensing, 7*(7), 9020–9044. https://doi.org/10.3390/rs70709020

Washaya, P., Balz, T., & Mohamadi, B. (2018). Coherence change-detection with Sentinel-1 for natural and anthropogenic disaster monitoring in urban areas. *Remote Sensing*, *10*(7), 1026. https://doi.org/10.3390/rs10071026

Wickham, H., Chang, W., Henry, L., Pedersen, T. L., Takahashi, K., Wilke, C., Woo, K., Yutani, H., Dunnington, D., & RStudio. (2022). *ggplot2: Create Elegant Data Visualisations Using the Grammar of Graphics* (R package version 3.4.0). https://cran.r-project.org/web/packages/ggplot2/

Wickham, H., François, R., Henry, L., Müller, K., & RStudio. (2021). *dplyr: A Grammar of Data Manipulation* (R package version 1.0.3). https://cran.r-project.org/web/packages/dplyr/

Wu, W., Shibasaki, R., Yang, P., Zhou, Q., & Tang, H. (2014). Remotely sensed estimation of cropland in china: a comparison of the maps derived from four global land cover datasets. *Canadian Journal of Remote Sensing*, *34*(5), 467–479. https://doi.org/10.5589/m08-059

Xie, Y., Allaire, J. J., & Grolemund, G. (2018). *R markdown: The definitive guide* (1st ed., p. 338). Chapman and Hall/CRC.

Xiong, J., Thenkabail, P., Tilton, J., Gumma, M., Teluguntla, P., Oliphant, A., Congalton, R., Yadav, K., & Gorelick, N. (2017). Nominal 30-m cropland extent map of continental Africa by integrating pixel-based and object-based algorithms using Sentinel-2 and Landsat-8 data on Google Earth Engine. *Remote Sensing*, *9*(10), 1065. https://doi.org/10.3390/rs9101065

Index